Unlearning Eugenics

George L. Mosse Series in Modern European Cultural and Intellectual History

Unlearning Eugenics

Sexuality, Reproduction, and Disability in Post-Nazi Europe

DAGMAR HERZOG

The University of Wisconsin Press

Publication of this book has been made possible, in part,
through support from the **George L. Mosse Program**
at the University of Wisconsin–Madison.

The University of Wisconsin Press
1930 Monroe Street, 3rd Floor
Madison, Wisconsin 537112059
uwpress.wisc.edu

3 Henrietta Street, Covent Garden
London WC2E 8LU, United Kingdom
eurospanbookstore.com

Printed in the United States of America

This book may be available in a digital edition.

Library of Congress Cataloging-in-Publication Data

Names: Herzog, Dagmar, 1961- author.
Title: Unlearning eugenics: sexuality, reproduction, and disability in post-Nazi Europe /
Dagmar Herzog.
Other titles: George L. Mosse series in modern European cultural and intellectual history.
Description: Madison, Wisconsin: The University of Wisconsin Press, [2018] |
Series: George L. Mosse series in modern European cultural and intellectual history
| Includes bibliographical references and index.
Identifiers: LCCN 2018011401 | ISBN 9780299319205 (cloth: alk. paper)
Subjects: LCSH: Eugenics—Europe—History. | People with disabilities—Civil rights—
Europe—History—20th century. | People with disabilities—Civil rights—Europe—
History—21st century. | People with disabilities—Nazi persecution. | Reproductive
rights—Europe—History—20th century. | Reproductive rights—Europe—
History—21st century. | Abortion—Europe—History—20th century.
Classification: LCC HQ755.5.E8 H47 2018 | DDC 363.9/2094—dc23
LC record available at https://lccn.loc.gov/2018011401

Contents

Illustrations

Unlearning Eugenics

Introduction

Few topics raise problems of precarious citizenship and dilemmas of moral argument and legal strategy more powerfully than the impasse currently evident across Europe (both Western and Eastern) between women's reproductive rights and disability rights. A major new tactic of activists and lawmakers, evident in a variety of guises depending on national context, has been to present the curtailment of abortion rights as a great advance for disability rights. Two traditionally progressive goals are thus turned against each other.

From Germany (2009), Spain (2012), and the United Kingdom (2013) to Hungary (2011) and Poland (2016), abortion opponents have begun promoting restrictions on sexual and reproductive self-determination as justice for the physically and cognitively disabled. Abortion on grounds of fetal anomaly has become one of the most fraught areas of bioethical and political controversy. The ethical complexities and practical consequences of this conflict have been immense. They are being played out in the courts and the media, in parliamentary inquiries and doctors' offices, and in countless private lives—even as austerity policies have meant that government support for self-determining lives for differently abled individuals is cut and the necessary conditions for flourishing lives are concretely constricted.

That the epicenter of the present impasse is Europe has an important prehistory in the Third Reich. The mass murder of the disabled in Nazi Germany anticipated the Holocaust of European Jewry, but its significance and effects in the postwar world have long remained unclear.

It was never self-evident what it might mean to unlearn eugenics. Indeed, in the first postwar decades, no one showed much interest in doing so.

Also among those Christian spokespeople, Protestant and Catholic, who styled themselves as offering moral direction to their societies in the nations of Western Europe and particularly in post-Nazi West Germany, early pride in having protested the so-called euthanasia murders was generally combined with the message that sexual conservatism needed to be restored and that above all abortion—on any grounds— must remain criminalized. And while Catholics held to their church's opposition to any interference with reproduction and hence rejected not only contraception and abortion but also sterilization, the government of West Germany found support among Protestants for its refusal to acknowledge that those coercively sterilized in the Third Reich should be understood as victims of Nazi "racial" policies. Some Protestants avidly worked to keep the Nazis' program of "eugenic" sterilizations analytically distinct from the "euthanasia" murders—specifically in order to advance their own postwar program of (supposedly voluntary) sterilizations under the rubric of "personal eugenics."

In all of this, individuals with disabilities, whether physical, emotional, or cognitive, were hardly ever understood or treated as agents with personal dignity and rights. Patronizing attitudes, ranging from pity to contempt, continued to be the norm; barriers to participation in political, social, and cultural life remained ubiquitous. In Germany, West and East, individuals who had been sterilized and family members of those murdered were treated as contaminated by the shame; popular support for the perpetrator doctors was demonstratively substantial. In numerous other nations—including, conspicuously, continuously democratic ones—eugenic ideas that had flourished with especial vigor in the first decades of the twentieth century persisted undisturbed by the facts of Nazism and its horrors and continued to inform treatment of individuals with cognitive disabilities and psychiatric conditions in particular.

It took a full four decades after the end of the war for radical disability rights movements—although forming in many Western nations in the mid-1970s as part of the wider efflorescence of countercultural protest—

to be catapulted into public view and mainstream popular and media discussion and to begin to be able to set the terms of debate about disability rights rather than being positioned as marginal and solely reactive. In the meantime, the sexual revolution sweeping Western Europe in the 1960s–1970s had transformed the conversation—and the laws—about abortion substantially. In this context, among the arguments put forward across Western nations, inherited and unquestioned negative attitudes about disability proved to be an unfortunately especially effective part of the moral armamentarium for advancing abortion rights. At the same time, sex rights activists in many countries, especially in the LGBT movements, had laid important groundwork for the later rise of disability rights.

One upshot of the complex and ever-evolving intertwining of issues of sexuality, reproduction, and disability was that while deprecatory notions of disability facilitated the decriminalization and broad popular acceptance of abortion in Western nations in the 1960s–1970s, by the 1980s–1990s the ascent of successful disability rights activism, due to an unexpected conjunction of dynamics that included enhancements in medical technology and growing popular hesitation about sexual freedoms, brought with it heightened concern about abortion on grounds of fetal anomaly. In addition, the collapse of Communism and the unification of Europe brought even more complications, as demography-preoccupied, fiercely nationalist, and aggressively sexually conservative versions of Christianity, both Catholic and Orthodox—hostile to abortion and homosexuality alike and positioning themselves against the purported licentiousness of the post–sexual revolution West—were suddenly in the ascendant. The terms in which the multidecade battles that had been fought in Western nations to get both women's rights to reproductive self-determination and the securing of dignity and comprehensive rights for individuals with disabilities understood as post-fascist imperatives were suddenly scrambled by the very different conclusions drawn in Eastern Bloc nations about what would constitute appropriate post-Communist lessons.

Many developments, then, have come together to alter dramatically the cultural landscape. Among the most important of these is the growing success of an energetically politicized postmodern religiosity in advancing its agendas in secular moral language, in tandem with the proliferating phenomenon—still far too little studied—of explicitly right-wing NGOs, many of them transnationally, even intercontinentally, funded and organized. Alliances and oppositions are reconfigured in new ways. Thus, for instance, while LGBT activism continues to provide important stimulus and conceptual frames for disability rights activism—not least with regard to sexual rights for individuals with disabilities, but also more generally with regard to pluralization of ideas about bodies, intimate interdependencies, and familial arrangements—right-wing religious and secular NGOs use their self-representation as opponents of eugenics, often also explicitly referencing the murders in the Nazi past, in order to restrict access to sexual and reproductive rights (including sex education, contraceptive access, and LGBT rights) for all citizens in the present. Ironically, then, the self-styling of many right-wing organizations as opponents of eugenics (while combining their hostility to new reproductive technologies with a reidealization of normative ideas of heterosexual coupledom and family life), along with the ongoing success of radical disability rights activists in defining eugenics as a form of racism (many of them more inclined to align with left-wing agendas), have both contributed to the growing resonance of the Nazi murders of the disabled as a touchstone in present political disputes (see figure 1). All of this,

Figure 1 (*right*). A drawing produced at one of the many workshops regularly held at the Hadamar memorial to the victims of the Nazi "euthanasia" crimes. Most of the picture is in gray, black, and white, including an image of one of the distinctive gray buses that took the victims to the killing centers, the now-iconic black-and-white checkerboard tiles on the floor of the basement "shower room" where the victims were murdered with carbon monoxide, and the foul black smoke billowing out of Hadamar's chimney from the crematorium into which the bodies were brought (in some cases after having the gold in their teeth extracted or their brains removed for research purposes). However, the contents of the thought-bubble are rendered in bright primary colors. It remains indicatively unclear whether the young woman whose vibrant but also fearful thoughts (a severed head, a surveilling eye) are going up in smoke is meant to represent one of the killing center's victims in 1940–41 or whether she is a visitor in the 2000s–2010s, empathically reimagining the horrors. Similar uncertainty surrounds the (colored in as blond

and blue-eyed) young man on the right. Is he a victim in the past or a recent visitor—or a perpetrator? Meanwhile, the man with closed eyes on the left could be a bystander from the past who is not wanting to acknowledge what was going on, although it was inescapably evident—or someone closing his eyes in the present, whether because of a lack of interest or in distress. Reprinted by permission of the Gedenkstätte Hadamar.

meanwhile, is happening in the midst of an inadequately understood—
but indisputably burgeoning—phenomenon of the "geopoliticization"
of sexuality, as high politics and diplomacy are imbricated with matters
of sex as never before and we find ourselves, globally, in a wider backlash against and ambivalence about sexual freedoms. It is this entangled
set of issues that the three chapters that follow begin to address.

My turn to the history of disability grew out of my work in Holocaust
studies, as well as in the history of sexuality and the history of religion—
although it was not immediately apparent with what extraordinary,
multifaceted complexity these three areas of research would soon inter-
twine. I had already begun transnationally comparative research on postwar disability rights activism and the evolution of the memorialization
of the mass murders (and had presented on this topic at a Holocaust
conference in 2008) when, in the midst of writing the book *Sexuality in
Europe: A Twentieth-Century History*, which eventually appeared in
2011, I ran across the source material that served as the initial basis for
the first chapter here. As I was collecting primary documents for the
chapter on the sexual revolution, I found a cluster of texts by Christian
theologians from the 1960s–1970s, both Protestant and Catholic, who
favored abortion rights—and who did so with innovative theological
arguments. This in itself seemed to me a remarkable find in view of the
rollback under way against women's rights to control their own reproductive lives being led, in the first decade of the 2000s, by religious
leaders from formerly Communist Eastern Bloc nations (just at that
moment—2004–7—being incorporated into the European Union)
and who were starting to make common cause with Pope Benedict XVI
and also with antiabortion activists in Western Europe. It was patently
apparent that a newly empowered phenomenon of conservative Roman
Catholics—part of a wider global revival of politically conservative forms
of religiosity causing observers in the early 2000s to begin talking about
a "postsecular" age and that included Jewish, Muslim, and Hindu trends
as well—was, together with evangelical Protestants and Orthodox
Christians, rewriting assumptions about what the content of Christian

faith and its implications for politics, sexual and otherwise, should be. This rewriting had a great deal to do with a long-unwinding backlash against the leftward turn of both Protestantism and Catholicism—a multiplicity of dynamics sometimes bundled under the rubric of "liberation theology"—that had done so much in the 1960s–1970s, the era of civil rights and anti–Vietnam War activism, to rescue postwar Christianity's reputation (in Europe well understood as besmirched by collaboration with numerous nations' fascisms) as a force for moral good.

Yet as I researched further the debates about abortion in the 1960s–1970s, I noted the plethora of references to disability—both among opponents of the legalization of abortion and among religious and secular proponents of liberalized abortion access. It turned out, indeed, as I soon discovered, that the debates of the 1960s–1970s over abortion rights in Western European nations had been saturated by references to disability, although this fact had not been incorporated into scholarship on the era. The Nazi mass murder of the disabled was an overt reference point—perhaps especially in France, where a commitment to *laïcité* (secularity) meant that religious arguments against abortion were less likely to be invoked by parliamentarians hostile to abortion rights. But the possibility of, or perceived need for, abortion on grounds of fetal disability—the so-called eugenic indication for abortion—also turned out to be a major focus of discussion. Many comments had a disdainful, unempathic tone, treating disability as a tragedy for families and a burden for societies. Simultaneously, I was noticing that in the early 2000s it was precisely the availability of abortion on grounds of disability, once one of the few grounds for abortion that had seemed acceptable, indeed immediately comprehensible, to broad majorities across the ideological and religious spectrum (in a survey conducted in 1971, for instance, fully 80 percent of Catholics in West Germany approved of abortion on grounds of fetal anomaly), that in the 2000s was being used by opponents of all abortions as a new entry point for regenerating a sense of moral conflictedness about abortion in general. In short, it was becoming unmistakable that unreflected insensitivities inherent in the prochoice rhetoric of the 1960s–1970s had come to haunt the abortion politics of the twenty-first century. That starting point ended up pulling me yet deeper into the riddle of the shifting lessons that were drawn from the Nazi past. But it also spurred me to search for inspirational resources from the pre- and post-Nazi eras, as well as to try to make better sense of the contradictory climate of the present.

As will become evident throughout, then, an important part of what makes this perhaps initially surprising conjunction of topics of distinctive interest for historians more generally has to do with what might be called the contrapuntal relationship between different moments in time. Or, to use other metaphors, what was always again notable in my research was the sometimes planned, but often quite uncontrollable, ricochets and repercussions between those past and present moments. Indeed, it would seem that at least with regard to the most contested issues of life, sex, and death, historical time simply does not accumulate or progress but rather reverberates in complicated ways. In addition, ongoing transformations in the ever-volatile present not only led me to pose new questions to the past but also brought me to recognize that my several objects of inquiry themselves quite apparently did not have stable meanings.

The chapters thus exist in some dialectical tension with each other, but that is both intentional and inevitable and arises from the intricacy and intensity of the material. The chapter sequence chosen here could be thought of as one of *thesis / antithesis / alternative antithesis*. The first chapter examines how negative attitudes about disability proved essential to advancing women's rights to reproductive self-determination in the 1960s–1970s across Western European nations—with, as noted, significant consequences for how opponents of abortion are succeeding strategically in the 2000s–2010s. The second chapter homes in specifically on West and then on united German discussions of the 1980s–1990s to explore how the (appallingly long-delayed) eruption of disability rights activism into mainstream politics and media discussion, coming as it did at the particular juncture in postwar history and retrospective engagement with the Nazi (and more immediate post-Nazi) past of the 1940s–1950s, caused—conversely to the developments of the 1960s–1970s—the new insistence on a more positive attitude toward disability to have adverse repercussions on women's rights to abortion access. The third chapter documents a range of disability rights efforts in the 2000s–2010s, widening out again to all of Europe, as it shows how activists have grappled with the philosophical heritage of the Enlightenment, revisited submerged strands within religious traditions, and utilized the (adapted) resources of psychoanalysis and poststructuralism in order to reconceive how we might talk more thoughtfully about such matters as sex, personhood, rights, autonomy, and self-determination. To add to the complexity of the subject matter, not only pregnancy but also disability itself are, after all, boundary-blurring phenomena that challenge long-cherished Enlightenment ideals of individual autonomy—even as the Enlightenment heritage remains indispensable for advancing the cause of disability rights in numerous realms.

Disability was a latecomer to the postwar human rights canon, and by turning attention to the non-abortion-focused disability rights trends of the 2000s–2010s, the final chapter provides yet another vantage point from which to see how activists have sought to undo the legacies of contempt and cruelty that the Third Reich exemplified in the extreme. Moreover, once again the complex reverberations between various presents and pasts are manifest as the final chapter traces a variety of literal and conceptual connections between avid present-day efforts to insist on full human subjecthood for the disabled, including especially the cognitively disabled, in all dimensions of existence and an eclectic but vital handful of intentional disabled-abled life-sharing communities founded in the 1940s, 1950s, and 1960s in either direct or indirect reaction to the National Socialists' systematic mass murder project. It also observes how significant the precursor developments of the sexual revolution and the LGBT rights movement of the 1960s, 1970s, 1980s, and 1990s have been for formulating a case for disability rights, expanding the postwar human rights lexicon and legislation, and finding imaginative ways to engender empathy and to communicate both the universally relevant implications and the minoritizing aspects of differently abled life.

It could be argued that disability has never been more accepted and less stigmatized than it is, thankfully, now. So while the present is marked by often brutal budget cutting, heightened mutual surveillance and erosion of privacy, and the frequent eruption of un–politically correct aggressions against the weak, these factors are continually jostling with and compensated for by an indisputably increased sensitivity to multicultural diversity and a—however ambiguous but nonetheless absolutely essential—rise in attunement to, even sentimental appreciation for, human vulnerabilities of all kinds. Yet in many of the recent popular and anxious debates about whether reproductive technologies and concomitant novel reproductive choices constitute a new form of eugenics—some critics aver that eugenics is now insidiously embedded within the liberal language of self-determination and individual risk management, while others insist that it is precisely the freedom of individual choice and the absence of state coercion that keep these new technologies and practices distinct from the pernicious eugenics of yore—very little attention has been paid to disability as it is actually lived. Since, in any event, the vast majority of disabilities only become evident after birth, one of the most important areas of inquiry in the present thus involves asking what it would mean to take disability rights seriously through the lifespan and to engage earnestly the genuine enormity of disability care and the intricacy of interpersonal imbrications—often lifelong—it can entail. It involves asking what emotional resources and political reconsiderations, as well as financial commitments, it would take in order to advance, concretely, the rights of individuals with cognitive disabilities in particular and to provide them and those who love and care for them with dignified rather than demeaned lives. It involves, in sum, refusing simple answers about what it would take truly to unlearn eugenics in post-Nazi Europe. That is part of what I begin to outline in the third chapter.

For multiple reasons, Down syndrome has become a particular flash point in the present debates. Among these are the obvious ones: its "visibility" in genetic material in the prenatal tests available since the early 1970s (hence often serving as the cause of a termination); its visibility (unlike autism or other forms of cognitive difference) in face and body after birth and throughout life; and, not least, its increasing visibility in the public sphere in the 2000s–2010s due to audacious and marvelous activism. But Down syndrome—precisely because of these multiple

Figure 2. Raoef Mamedov, *The Last Supper* (1998). Courtesy of the Lilja Zakirova Gallery, the Netherlands.

visibilities—is often also instrumentalized by abortion opponents with romanticized images that deny the broad variability of outcomes and minimize the stark realities of lifelong care. As I was researching, and as the early twenty-first century kept unfolding, I ran across the photographic artwork of the Moscow-based Azerbaijani filmmaker Raoef Mamedov and was moved especially by his *Last Supper* (1998), a remake of Leonardo da Vinci's version but with Jesus and the disciples' roles performed by individuals endowed with Down syndrome (see figure 2). Was this provocative and exploitative? Or was it in fact extraordinarily theologically profound?

Mamedov has created numerous photographic versions of famous Christian artworks employing individuals with Down syndrome, many of them his close friends, as well as other remakes of paintings involving patients of psychiatric clinics as his models and subjects, often acting together with them in the scene in order to generate a particular mood. Mamedov had previously worked as a nurse in a psychiatric hospital and was strongly influenced by the "schizoanalysis" of French philosopher Gilles Deleuze and psychoanalyst Félix Guattari, coauthors of the 1972 counterculture classic *Anti-Oedipus* (as it happens, and as will be discussed in the third chapter, an oft-cited reference point for both queer theorists and disability theorists in the early twenty-first century), as well as by the intimate and vulnerable style of psychoanalysis pioneered by Sigmund Freud's longtime associate Sándor Ferenczi. Yet could the Jesus with Down syndrome also be seen from the vantage of a concurrently developing project growing out of the global liberation theology movements of the 1960s–1970s, which at various times had imagined God as black, as red (i.e., Native American), or as a woman and which is now again evident in the vibrant expansion of a theology of disability, one of the most active areas within disability studies more generally? Nancy Eiesland in *The Disabled God* (1994) may have been the first deliberately to think through the image of the Christian God from the perspective of disability, calling attention to the significance of a Savior with a wounded and broken body but also imagining God memorably as a being in a "sip-puff wheelchair"—the kind used by quadriplegics who are able to move the wheels through sipping and blowing through a straw. Since then her pioneering work has been amplified and extended into discussions of theology in relation to cognitive disability, most impressively by Amos Yong (*Theology and Down Syndrome*, 2007) and John Swinton (*Dementia*, 2012, and *Becoming Friends of Time*, 2016). But only Mamedov has been radical enough in his artwork to force viewers to envision the Christian God himself as a being with a cognitive disability. There is nothing sentimental or romanticized here. The altered perspective—precisely because it preemptively defies being rejected as heretical or sacrilegious, lest anyone who were to accuse it of being so reveal themselves to be impiously denigrating of disability—is brilliantly salutary.

"Religion," I will be emphasizing recurrently, does not suffice as an appropriate shorthand to explain the widening resistance worldwide to women's self-determination in general and to parents' ability to make—often deeply anguishing—personal decisions about terminating pregnancies in particular. What is it, then, that makes restrictions on sexual freedom and reproductive self-determination acquire such powerful popular appeal—and this across so many different cultures—in the early twenty-first century? What might that have to do with everyone's sense of growing precarity in a world of massively exacerbated inequities? Why are intimate rights of all kinds so hard eloquently to defend? How did we get to the current impasses? These questions remain.

1

Abortion and Disability

Western Europe, 1960s–1970s

The legalization of abortion is a step on the path of a monstrous regression. When one forgets that the right to life is inviolable one can, after introducing abortion, also recommend actions against the physically or mentally handicapped, against "useless eaters," against the incurable, against those who weigh heavy on the society and thereby arrive, my dear colleagues, at the worst Nazi racism.

deputy to the French National Assembly René Feït, 1974

It would be not only grotesque, but actually blasphemy, if one tried to give the responsibility for a rape, for clumsiness, or for technical failure to that God whom Jesus proclaimed as the good Father of all peoples.

Swiss theologian Gyula Barczay, 1974

Abortion was and is a topic that cuts across the (only ostensible) rift between religion and secularism while also revealing the complexities and mutability of each. To reimmerse oneself in the specifics of religious and secular arguments over the decriminalization of abortion in Western European nations in the 1960s–1970s is to realize fairly quickly

15

that few things were as we remember them now. The master narratives generally circulating in scholarship about the 1960s–1970s in Europe—about the decline of religiosity and the rise of secularism, about the sexual revolution and the ascent of feminism—have almost all missed just how much difficulty abortion rights activists actually had in developing moral arguments for legalizing the termination of unwanted pregnancies. They have missed as well two other matters of singular importance. One has to do with a remarkable efflorescence of efforts at the time to make a case for abortion rights *within* a religious, specifically a Christian, framework. The other has to do with the—in hindsight disturbing—striking preponderance of a reliance among both religious and secular proponents of expanded abortion access on the assumption that to bear and raise a disabled child would be an especially awful fate. For as it turns out, and quite unexpectedly in view of the extant master narratives, references to disability saturated the abortion debates of the 1960s–1970s. This can be read *both* as an indication of the persistence of contempt for the disabled in the post-Nazi era—and this not just in formerly fascist nations but also in neighboring, continuously democratic ones—*and* as a telling sign of what acute ambivalences surrounded the sexual revolution of those decades and of just how near impossible it was, apparently, to argue straightforwardly for women's rights to sexual pleasure without reproductive consequences. Abortion quite evidently was never just about itself. The topic always brought with it a jumble of associations—often involving inchoate but deeply held feelings about femininity and motherhood, about sexual practices and pleasures, and about demography and eugenics.

Reading the arguments of theologians, parliamentarians, journalists, and feminists with and against each other makes it possible to reconstruct previously neglected aspects of the debates that took place in Britain, France, West Germany, Italy, and Switzerland in the 1960s–1970s. A multinational comparison is especially useful for sorting out the similarities and differences in the arguments and political strategies used by all sides, as well as for noticing the transnational flows across borders among both proponents and opponents of abortion. The comparison of these particular five nations is revelatory not only because it permits a juxtaposition of Catholic and Protestant and mixed-confessional countries, in four of which liberalization of the law succeeded in the 1960s–1970s and in one (Switzerland) in which it failed. It also helpfully highlights the contrasts between nations with a fascist or collaborationist past (Germany, Italy, France) and those that were uninterruptedly democratic (Britain and Switzerland). The legacies of Nazism turned out to be transnationally relevant—even as the meaning of those legacies was deeply contested.

At the same time, it was by no means then, in the 1960s–1970s, nor is it at all now, as we head toward the end of the second decade of the twenty-first century, in any way obvious what exactly should count as secularity and what should be considered a form of religious renewal. In fact, precisely the 1960s–1970s were a time of the most ardent dispute, among theologians and laity alike, about the lessons of the Gospels, the relationship between faith and politics, and the very nature of God. In those years, for example, an increasing chorus of Christian theologians argued that God should no longer be thought of as some kind of supernatural magician moving people around like chess pieces, permitting or averting plane crashes or mining disasters and responding to anxious prayers by working miracles. This should rather be thought of as superstition and not faith. Others, relatedly, argued that those who used religion to support such hideously immoral enterprises as the Vietnam War and consumer-capitalist materialism based in massive economic inequities, *those* should be considered the secularizers. By contrast, these commentators suggested, those who instead understood God as powerless, not powerful, as suffering with human beings when they suffer, and as needing human action to improve the world, *those* were the religious renewers, the truly faithful. Sincere faith, it was contended, was not about maintaining stale dogma but rather about orthopraxis in the world. Particularly in the West German context, these kinds of theological arguments were very much bound up with efforts to draw lessons from the Nazi past—and to undo the impact of the conservative and apologist forms of Christianity that had predominated in the first post-Nazi decades. But also in Britain, France, Italy, and Switzerland, numerous theologians and activist laity, Protestant and Catholic alike, under the impact of *nouvelle théologie*, personalism, Vatican II, the Death of God movement, political theology, or liberation theology, argued that the churches should adapt their messages to the changing times and practical human needs, that older notions of God needed revision, and that solidarity with the oppressed was the way to be properly faithful.[1]

These debates about the nature of God and the demands of faith and the relationship between religion and politics were complicated yet further by arguments over sexuality and reproduction. All through the first three-quarters of the twentieth century, there had been a tight link between sexual politics and religious politics. On the one hand, nothing so turned people off from religion in general as being told what to do (or what not to do) in bed—even as this dynamic played out in quite diverse ways. Clergymen and theologians were acutely aware that matters of sexual pleasure and fertility control were at the core of their parishioners' daily concerns, with reactions ranging from severe internal conflict over the use of birth control or preference for certain sexual practices, to various—inevitably somewhat mendacious—compromise formations between church teachings and personal behaviors, to complete alienation from the churches. However, the intertwined histories of sex and religion were yet more complicated, for, on the other hand, we find also that there were recurrent sex-liberal movements among the Christian clergy. For instance, there were French, Dutch, Belgian, Italian, and Irish priests in the 1930s–1960s who supported their parishioners' desperate efforts to reconcile faith and family planning.[2] The Belgian cardinal Leo Jozef Suenens, archbishop of Malines-Brussels, as well as the Roman Catholic bishops of West Germany, Austria, and Switzerland even challenged directly the Vatican's stance on the birth control pill.[3] There were Dutch and French and British clergymen and prominent laypeople active in homophile rights movements in the 1950s–1960s.[4] And, as will be discussed in detail below, there would be both Catholic and Protestant clergymen and theologians in every Western European nation in the 1960s–1970s who argued in favor of legal access to abortion as not just a lesser evil but an explicitly Christian desideratum.

Yet when we look at the twenty-first-century present, especially in post-Communist Eastern Europe, albeit with increasing frequency

Figure 3. Antiabortion demonstrators in Warsaw, June 2011, with banners of bloody fetuses and Adolf Hitler as abortionist. In the foreground, a feminist counterdemonstrator raises a handmade poster declaring "LIFE FOR FETUSES!!! DEATH TO WOMEN," since this was how feminists summed up the new law proposed in Poland at that moment that would have banned all abortions without exception. The law ended up failing in the parliament by only five votes. Feminists involved in the counterdemonstration found that it took the antiabortion demonstrators some time to figure out that the woman was not one of them. Reprinted by permission of Zofia Nawrocka, Porozumienie Kobiet 8 Marca, Inicjatywa 8 Marca.

also across the West, we often find religion functioning in a dramatically opposed way. In the post-Communist nations of the European Union, the revival of an assertively nationalist, demography-preoccupied, anti-Western Christianity (whether Catholic or Orthodox) is very frequently used to justify new—indeed, rather postmodern, not at all only traditional—forms of hostility both to homosexuality and to abortion (see figure 3).[5] And within Western Europe, novel forms of politically and sexually conservative Christianity (both Roman Catholic and evangelical Protestant), as well as new sexually conservative secular movements, are utilizing the courts, the media (including the internet), and mass public demonstrations in order to challenge sexual liberalism and curtail access to abortion rights. (For instance, in Italy in 2008, antiabortion activists argued that the UN campaign for a moratorium on the death penalty should be emulated by a moratorium on abortions and that abortion clinics should be emblazoned with the Auschwitz-echoing slogan "abortion makes you free," while in Belgium in 2010, both secular and religious abortion opponents borrowed from feminist rhetoric to contend that women are victims of "abortion culture" and that "women deserve better than abortion."[6]) It is not least, then, the twenty-first-century resurgence of religious arguments against abortion that makes it worth recovering the reasoning of theologians and activist Christian laity who, in the 1960s–1970s, actively favored expanded abortion access. Nonetheless, as noted at the outset, in the midst of all the strong arguments made at the time in favor of liberalized abortion access, there was also evident an unreflected insensitivity to the disabled—and it is this that is now, in the twenty-first century, being turned against women's rights activists.

Open Secrets

In order to make sense of the debates of the 1960s–1970s, however, the first task is to consider the numerous *open secrets* surrounding the topic in those decades. One important point is that while there was certainly a correlation at the time between the strength of Catholicism in a nation and the difficulty of undoing the laws that criminalized both contraception and abortion, there was also a correlation between the strength of Catholicism in a nation and the extent of reliance on abortion as a fertility management strategy. In other words, the more Catholic, the more babies. But additionally, the more Catholic, the more abortions.[7]

A second, and related, open secret of the era was that the illegality of abortion had at the time not so much to do specifically with protection of "life," as was claimed by religious conservatives then and as has been retrospectively assumed by historians; instead, it was part of a broader preexisting pattern of hostility also to contraception. The laws against the promotion and sale of contraceptives promulgated in 1920 in France, 1930 in Fascist Italy, and 1941 in Nazi Germany were about demographic and national strength and had nothing to do with the sanctity of life per se. These were the laws still on the books in the 1960s. Meanwhile, what bears emphasis is the persistence of concern with demography and ongoing, widespread ambivalence about contraception also in the post-Fascist, post-Nazi, post-Vichy era.[8] From Italy to West Germany to France, legislators and courts continued to restrict contraception and to worry about national birthrates. Indeed, in 1946 the Italian government restricted the circulation of information about contraception even further than it had under Fascism.[9] And as one West German court put it—defensively but assertively—in a 1955 decision: "Not every legal measure that serves population growth has a National Socialist tendency. . . . For every healthy state . . . a growth in population is absolutely desirable."[10] This ambivalence about contraception and interest in raising birthrates among those deemed "healthy" (i.e., nondisabled) was clearly evident in continuously democratic nations like Britain and Switzerland as well. The influential Swiss Protestant advice writer Theodor Bovet spoke for many when he insisted in 1955 on the need to "be concerned with the healthy inheritance of our *Volk*" as he bemoaned the fact that "the less valuable elements, especially the mentally deficient, reproduce themselves approximately twice as much as healthy families. It is therefore absolutely necessary that, if we do not one day want to be completely flooded by those [elements], that everyone who feels himself to be healthy . . . give life to as many children as possible."[11]

What remains unclear—and perhaps this very lack of clarity can give us a valuable insight—is whether this resistance to contraception was "really" about demography or "really" about a certain view of what women were for. For perhaps it was both. In the 1920s–1940s, declining birthrates had been interpreted in every nation as a weakening of the nation's strength. But it was definitely not irrelevant that women's inability to control their own reproductive life choices made them dependent and vulnerable by definition. Moreover, ambivalence about women's freedom clearly persisted into the era of the sexual revolution. Half of the young British men polled in a 1970s survey objected to the birth control pill because it gave women sexual freedom, experienced as a "threat" to the men's "dominant role."[12] Italian sex rights activist

Luigi De Marchi, a pill proponent, observed that it was *men* who were the ones resistant to contraception: "They worried that without the fear of pregnancy their wives would 'be free to go with someone else.'"[13] Or as Simone Veil, the French minister of health in Valéry Giscard d'Estaing's cabinet, reflected retrospectively on the ugliness that was spewed at her during the 1974 parliamentary debates as she presented the bill that would suspend criminalization of abortion for the following five years, already since the legalization of contraception in France in 1967 and then even more with abortion in 1974: "The men were afraid that the women were slipping away from them" (*les hommes avaient peur que les femmes leur échappent*).[14]

For women, the nonavailability of contraception was devastating; sex, also within marriage, was for hundreds of thousands of women so often a focus of fear and not of desire. Contraception itself was a source of tremendous awkwardness. As one West German family planning expert remarked, there was in the populace enormous ambivalence about "the incursion of reason into the realm of the sexual."[15] One female pharmacist who wrote to Simone de Beauvoir after *The Second Sex* had been published described the many women who begged her for advice but then ended up relying on illegal abortion and for whom the lack of contraception made sexual pleasure impossible. To emphasize the point, she described one of her customers, a twenty-nine-year-old woman aged far beyond her years, for whom the husband's days off from work were "torture." She had had four children, ten miscarriages, and "no pleasure, ever." Many of the women this pharmacist knew were like "beasts caught in a trap": "They do not dare demand in public what they weep for in private."[16]

Behind all the glaring open secrets, in short, there were yet more. They concerned sexual habits; intimacy and tension; longings (often unfulfilled) surrounding sex and the discomfort felt by many in talking explicitly even with a spouse or lover; the desire for particular experiences of pleasure (especially intense, or especially uncomplicated); the subjection of women as part of what made sex erotic for some men; the assumption also among women that femininity was inevitably about masochism, caring for others' needs, and self-sacrificial devotion; and both men's and women's conflicted feelings about reproduction and the value of motherhood.

In general, though, the greatest and most obvious open secret of European sexual cultures in the 1960s was the prevalence of illegal abortion. Estimates for tiny Switzerland ranged between twenty thousand and fifty thousand illegal abortions a year, in addition to the more than twenty-one thousand legal ones undertaken under the rubric of the "maternal health indication," in place since 1942. Estimates for Britain ranged from forty thousand to one hundred thousand per annum. The other nations' rates were far higher. Estimates for West Germany assumed one million abortions every year—one for every birth. Estimates for France ranged between three hundred thousand and one million a year. Estimates for Italy ranged between eight hundred thousand and three million a year.[17]

This background helps explain why the eventual success of decriminalization in all three of the latter countries—in France in 1974, in West Germany partially in 1976, in Italy fully in 1978—was based not on the feminist slogan "my body belongs to me" but rather largely on abortion rights advocates' and supportive lawmakers' emphasis that *the law itself had lost both force and respect by being so widely disobeyed*. Simone Veil said it most bluntly as she presented her bill to the French parliamentarians: "We have arrived at a point at which the authorities can no longer evade their responsibility. The current situation is awful, lamentable, indeed dramatically so, because the existing law is being openly mocked, in fact ridiculed. We are in a situation of disorder and anarchy that can no longer be sustained."[18] In West Germany, the newsmagazine *Der Spiegel*—a strong advocate for abortion rights—made the case that "there is hardly another law in the Federal Republic which is so routinely flaunted as the one against abortion—every day more than a thousand times."[19] And in Italy, as the feminist group Rivolta Femminile put it in 1971, "We . . . insist that the one to three million secret abortions that are estimated to be occurring in Italy every year are enough to make the law that criminalizes abortion de facto invalid."[20]

This—at once pragmatic and earnest—argument that the law was being circumvented daily would be a very strong factor pushing parliamentarians in each nation toward decriminalization. What helped as well was the sense felt by many government leaders that they needed to become more woman-friendly in their legislation; the times felt ripe for reform. From Giscard d'Estaing's deliberate effort to style his government as more female-friendly than the more conservative government of Georges Pompidou that had preceded him, to West German Social Democratic Minister of Justice Gerhard Jahn's determination to advance policies supportive of women, to Italian political parties' mad scramble to provide possible abortion legislation once they realized how strong the popular demand for liberalization was, centrist government leaders understood that the status quo was no longer acceptable.

Ultimately, of course, the pressure of masses of tens of thousands of women (and supportive men) protesting in the streets would be absolutely essential in shifting what politicians could find imaginable (see figure 4). So too were a plethora of other inventive activist initiatives. These included public self-accusation campaigns by women who had had abortions or by physicians who had performed them; deliberately publicized collective travel to abortion providers in other nations where the service was legal or where (e.g., in the Netherlands) authorities had simply stopped prosecuting abortion providers; clinics and organizations that openly announced they would provide abortions; and deliberately dramatized court cases.[21] It eventually took a complex mix of courage to put oneself at risk, lawyers' strategic use of scandal, opinion surveys demonstrating the growing breadth of popular support for decriminalization, and, not least, sharpening splits among conservatives over sexual politics for the so widely disobeyed laws actually to be changed. Over and over, the open secrets had to be made even more fully open as the blatant contradictions and hypocrisies structuring the various national cultures' handling of abortion were exposed and debated.

Among the most important arguments put forward in all countries were those having to do with the devastating damage to women's health caused by the many illegal abortions, while support from left-wing parties was additionally secured through the argument that class injustice was pervasive in a situation in which women of means could travel abroad or pay for private doctors to secure abortions while poorer women were

Figure 4. Abortion rights protest in Rome, April 1977. Photo © Paola Agosti.

vulnerable to butchery by opportunistic quacks. It was pointed out that thousands of doctors made millions in untaxed income by secretly offering expensive illegal abortion services on the side—and that it was reluctance to lose this income that caused them to oppose legalization. The West German humor magazine *Pardon* made clear that medical doctors, like their quack competition, were personally invested in keeping abortion illegal and their income from it hence untaxable.[22] Or as the pro–abortion rights British *Guardian* warned sarcastically, any reform of the law in the UK would "cramp the style of such easy-going doctors."[23] Elite doctors could make thousands of pounds sterling, while "abortionists from humbler streets are able to perform perhaps 100,000 illegal operations every year, killing a few of their patients, rendering many permanently sterile, and exposing all to experiences which, if suffered by men, or horses, would long ago have engaged the ingenuity of parliamentary reformers."[24]

All through the years of struggle, however, there remained the nagging and urgently felt need to engage the debate on religious grounds directly rather than trying to evade that terrain. Protestants and Catholics fought not just for contraception but also expanded abortion access, making their cases for either the "trimester model" (abortions available for any reason in the first trimester) or the "indication model" (either three indications—medical/maternal, criminal, or eugenic—or four indications, including also the socioeconomic).

Religious Arguments

One of the principal arguments made centered on the emphatic idea that moral reasoning on the subject of abortion needed to begin from the life of the individual woman in her specific situation—and that her life needed to be considered in its holistic entirety. As the Methodists in the UK noted in 1966, "The most important fact about a woman seeking an abortion is not that she is about to commit a crime, but that she is a human being in need."[25] In response to the question posed to Jesus, "Who is my neighbor?" the answer—according to West German, French, and Swiss theologians—had to be first and foremost the woman herself.[26] They argued that the much-invoked idea of "reverence for life" was in this context an abstraction not in keeping with Jesus's teaching that "loving one's neighbor" needed to start from the position of the already living involuntarily pregnant woman and her sense of dread about the future. Not the "abstraction 'human life'" but the "concrete human being," argued the Swiss Protestant theologian Gyula Barczay, should be the starting point for moral reflection.[27] The already living woman deserved absolute priority. Moreover, health too needed to be broadly understood. In the discussions preceding the Abortion Act, which would be passed in the UK in 1967, a commission of the Anglican Church went on record with its overall conclusion that "in certain circumstances abortion can be justified"—not only when "it could be reasonably established that there was threat to the mother's life or wellbeing" but also because her "health and wellbeing must be seen as integrally connected with the life and wellbeing of her family."[28] It was shameful that somehow a woman's already existing life, in all its richness and complexity, was treated as equivalent or even of less value and as being in competition with the incipient life she carried.

Second, theologians argued that it was a foundational moral issue to be able to determine the number and timing of children, that this was a rudimentary part of human dignity. Many commentators noted simply that it was a moral positive that women were being increasingly treated with greater respect and dignity, that they were able to take part in public life and to work outside the home, and that reproductive self-determination was itself a moral value. The denial of this aspect of human dignity was a crime, a repugnant form of disrespect for women. For Italy in particular, framing the abortion issue as one of social justice more than individual liberties helped substantially in making passage of the law that decriminalized abortion possible.[29]

Third, interestingly, a number of theological commentators argued that *wantedness* and *relationality* were central aspects of what constituted humanness in the first place, and they vigorously questioned whether it was moral to force a woman to bear an unwanted child. A team of French Catholic physicians, social scientists, and theologians published in the Jesuit journal *Études* provocatively suggested that an incipient life needed to be affirmed—"called to be born"—in order to be fully humanized; an abortion, they contended, was "not murder, because it is specifically motivated by the refusal or incapacity to humanize the embryo." Termination of pregnancy, the authors proposed, was morally justified precisely when its aim was to prevent the dehumanization of an unwanted child.[30] Related arguments were put forward by West German Protestants and by Swiss Protestants and Catholics: "Not the physiological conception, but only human acceptance makes life as human life possible."[31] And "It's a matter, in short, not of some opposition between the 'right to life' and the 'right to be wanted'; it's a matter of the realization that 'wantedness' is a foundational condition of the humane quality of human life and that this condition cannot be forced via the threat of punishment. The question 'May it live?' must be counterposed by, with full equivalence, the question '*Must* it live?'"[32]

Meanwhile, and fourth, although sex in its intricacies was not discussed by the theologians, some did acknowledge that—as the authors in *Études* remarked and as the Swiss German Catholic theologian Stefan Pfürtner, a member of the Dominican order, appreciatively repeated—it was just and right that "women will no longer accept it that they are the ones who have to pay for the pleasure of the men, especially of those men who do not concern themselves with the possible consequences of sexual relationships."[33]

Fifth, and importantly, many theologians worked strenuously to reject the idea that life begins at conception, insisting that the potential life of the embryo (or as the British Anglicans put it, "this still unformed human organism") should not be confused with the human life of a later-term fetus.[34] The Catholic Pfürtner was especially adamant in emphasizing the signal significance of this distinction. He endorsed the idea that "the fetus—especially in the first weeks—should not be classified as an independent human being, and the termination of pregnancy should be understood as merely an intervention in the bodily life of the woman." The fetus was "a biological something" but not yet a human life.[35] And the Anglican commission expressly criticized the idea that there could be any certainty about whether an embryo could be said to have "a living soul."[36]

The sixth and most notable argument, however, was the one that directly challenged the idea that all pregnancies were God's will. The Swiss Protestant theologian Gyula Barczay was especially forceful in repudiating the idea that God demanded all pregnancies to be continued. That was pure biologism, he said, not true faith. (In short, Barczay directly reversed what counted as secular and what counted as an act of faith.) To treat God as the origin of a conception that was caused by rape, male ineptitude or irresponsibility, or technical failure (like a slipped condom or diaphragm) was, in Barczay's eyes, "not only grotesque but actually blasphemy" and simply incompatible with Jesus's teachings.[37]

Bringing in Disability

Finally, though, what stands out most now in rereading the texts from the 1960s–1970s is how prevalent references to disability were. To legalize abortion, West German Catholic opponents of abortion had contended in the 1970s, would be "the most disturbing attack on the moral foundations of our society since 1945" and "the largest Auschwitz in European history."[38] In addition, they expressly invoked the Nazi "brown thugs" and their "murder of the cripples and the sick."[39] French parliamentarians in the 1970s who were opposed to decriminalization also invoked "Nazi doctors," "genocide," "racial eugenics," "organized barbarism protected by law, as it was, alas! thirty years ago by Nazism in Germany," "crimes perpetrated during the last war," and "crematoria fires." And they argued that to legalize abortion would be the first step in a "monstrous regression" that would lead to the euthanasia of the handicapped, the murder of "useless eaters" (*bouches inutiles*).[40] (It is notable that in *laïcité*-oriented France, religious arguments were hardly used by politicians. Instead, references to the horrors of Nazism fulfilled the moral function.)

But both secular and religious advocates *for* decriminalization also invoked disability recurrently.[41] Britain was the first country in Western Europe outside of Scandinavia fully to decriminalize abortion—already in 1967, really a prefeminist moment—and there is no question that this was not just due to the hope of reducing the incidence of illegal abortions and the damages they did to women's health but also to the early 1960s scandal of thousands of birth defects caused by the morning sickness pill thalidomide. Not until 2012, fifty years too late, did the pharmaceutical company the Grünenthal Group formally albeit inadequately apologize and establish a memorial in Stolberg, Germany.[42] A thousand children in Britain (at least ten thousand worldwide) had been born with truncated limbs. And although this is less well known, a further thousand children in Britain had died within a few months of birth because the drug could additionally cause malformed inner organs.[43] Many of the women carrying these children had sought abortions but been denied them. Additional scandal surrounded birth defects such as brain or heart damage or hearing loss caused when the pregnant woman had a case of the disease rubella. But it was especially the scandal surrounding the deaths of children with malformed organs that made the inclusion of the eugenic indication in the 1967 Abortion Act appear to be imperative and self-evidently moral.[44] Already for years in the run-up to the 1967 decision, the British press had carried articles expressing outrage specifically at the lack of availability of abortion in cases of fetal disability (and this despite the fact that the relevant ultrasound technology was just being developed, so that only known toxins or illnesses could predict an outcome of disability). The Anglican commission, in its reflections in 1965, spent extensive time arguing that the thalidomide case and other cases of anticipated fetal abnormality or deformity could certainly make abortion be the *moral* choice—especially in view of a couple's anxieties about their capacities to raise a disabled child effectively. In Italy, disability would become a factor as well, though on a far smaller scale—due to a chemical factory explosion in Seveso near Milan. Many women had sought abortions because they feared fetal damage.[45]

But eugenics would come to factor in the discussion of abortion in multiple and not only ethically sensitive ways. *Der Spiegel* in its pro-abortion writing in 1971 expressed glib disgust that 15 percent of West German medical doctors had argued that "children should be born against the will of the mother, even if they will come into the world as cripples or mental idiots [*Krüppel oder Schwachsinnige*]."[46] The French team writing in *Études* opined—with remarkable tactlessness—that it was actually immoral for those children to be forced to be born who would end up being a "heavy burden" to society.[47] Less crassly, and with more anguish, West German and Swiss theologians emphasized the importance of compassion for women requesting abortion due to concern about fetal disability.[48] While recognizing that sometimes, for instance, a Down syndrome child (at the time referred to as "mongoloid") could bring great love and joy to its parents (noting that Down syndrome children did have "a happy consciousness") and while expressing worry about sliding "onto the slippery slope to the killing of so-called 'life unworthy of life,'" they nonetheless emphasized the extraordinary weight that a disabled child could become for the mother; for her marriage, especially if it was labile already; for the surrounding society; indeed, for itself.[49]

Much can be said about this important phenomenon. One point, as noted, is that it is certainly a sign of how difficult it apparently was for defenders of abortion rights to articulate their case forthrightly. A similar sign of that difficulty were the many gestures that were made at the time to the purported danger of global overpopulation as somehow a significant moral justification for the use of the birth control pill within the West.[50]

But several further points need to be made. The first is that eugenic argumentation was part of the history of battles for contraception and abortion from the very beginning of the twentieth century on—and it was racist in its inception: in its condescension toward the lower classes within Europe; in its worry that the brown, black, and yellow peoples of the world were "outbreeding" the white peoples; and in its contempt for the disabled. What becomes—disturbingly but revealingly—clear when one reads procontraception arguments from the 1910s–1930s is how completely eugenic assumptions saturated the common sense of the era.[51] It was in those decades more difficult—for many, apparently impossible—to argue for women's rights to sexual pleasure than it was to use denigration of the disabled as a seemingly moral argument for the value even of contraception. This could partially be seen as sympathetic to poor women and the damage done to their bodies by repeated pregnancies and illegal abortions. It was certainly, for instance, what motivated the German gynecologist Wilhelm Mensinga of Flensburg, son of a pastor and himself a believing Christian, who invented the diaphragm—and who also recommended abortifacient strategies if contraceptives failed. And it is what motivated the Dutch contraception activist Aletta Jacobs, the first female doctor in the Netherlands, who promoted the Mensinga pessary among the Dutch working classes. But the defense of contraception could be expressed in very ugly terms. In the early twentieth century, the Swiss physician Auguste Forel was especially blunt: "The sick, the incapable, the mentally deficient, the bad ones, the inferior races must be systematically educated to birth control. The robust, good, healthy and mentally higher standing ones, however, must be, as I have repeatedly argued here, encouraged to multiply strongly."[52] In the early 1930s the Spanish socialist and feminist sex radical Hildegart Rodríguez advocated for legislation that would allow women to prevent the birth of children who were "retarded, epileptic, degenerate, insane."[53]

Moreover, eugenic argumentation against the lower classes and against the disabled continued *after* Nazism was defeated—also in continuously democratic nations. Inquiries in postwar Britain about views on contraception recurrently triggered responses that revealed the ongoing significance of eugenic attitudes and utter lack of self-consciousness in expressing them. One woman, mistress of a school, in response to a query about "your attitude to birth control," spontaneously offered this: "Unaesthetic, but probably necessary in many cases. Proper social training should deter physically-unfits from having children. Mentally-deficients should be prevented if necessary." And a schoolmaster opined that birth control was "being used by the wrong people. Intelligent people should procreate and give us more of their sort. But it is the semi-morons who breed like rabbits."[54]

It was, in short, apparently quite hard to unlearn eugenic thinking.[55] It is indeed an enormous achievement for justice and human rights that disability rights are now at long last on the agenda not just of activist organizations but of European governments and the European Union as well, and extraordinary—although still inadequate—progress has been made just in the last fifteen years.[56] But just as disability rights have gained a belated—although still too fragile—hold on public consciousness, they are almost instantly being pitted against women's rights to abortion access. Antiabortion activists are specifically presenting the availability of abortions on the grounds of fetal disability (and now also the possibility within IVF not to implant a fertilized egg with Trisomy 21 or another genetic condition) as in and of itself a horrendous affront to disability rights. Reproductive rights activists are very much on the defensive.[57]

Recent Trends

In Germany in 2009, for example, the law was changed to intensify restrictions on abortions undergone on grounds of fetal disability. Christian Democrats had attempted to implement these restrictions already in 2001 and again in 2004; the change succeeded in 2009 in winning the support of Social Democrats and Greens because the issue was framed as one that advanced the cause of disability rights.[58] In 2013 the British Parliament held the "Inquiry into Abortion on the Grounds of Disability" to investigate whether the contrast between the extended deadline for terminations on grounds of the "eugenic indication" when compared with the earlier-in-pregnancy deadline for abortions on other grounds constituted, in and of itself, an injustice to individuals with disabilities and hence a violation of nondiscrimination legislation in the form of the Equality Act of 2010.[59] In Spain in 2013 the ruling conservative Partido Popular strove to reverse the freshly (in 2010) achieved liberalization of abortion law, in part on the grounds that Spain needed to come into compliance with the United Nations Convention on the Rights of Persons with Disabilities (adopted in 2006). In the Partido Popular's proposed law, abortions on grounds of disability would be only allowed under the condition that the "anomalies" or "deformities" are "anomalies incompatible with life," which would have limited women's access to abortion even more than it had been under the restrictive 1985 Spanish law. Alberto Ruiz Gallardón, Spain's minister of justice and author of the bill, had declared already in 2012 that it would be "ethically inconceivable" to provide an "unborn baby" with "any kind of disability or malformation" reduced protection.[60]

Also in formerly Communist nations like Hungary and Poland the anti-"eugenics" argument against abortions has gained momentum—in the case of Hungary in the form of articles inserted in the Constitution formulated in 2011 and promulgated in 2012; in the case of Poland articulated by none other than the prime minister, Beata Szydlo, herself. Already in the more immediate aftermath of the collapse of Communism, scholars pointed out that in many former nations of the Eastern Bloc, nativist pride and antagonism toward Western culture were manifesting in the form of vociferous demands for a return to conservative notions of gender roles and that hostility to abortion was fueled both by the strong emotional association between legal abortion and the Communist past *and* by deliberate fanning of demographic anxieties.[61] The addition of concern with "eugenics" and disability is far newer.[62]

The Hungarian Constitution blends invocations of the nation's Christian tradition with declared opposition to both Nazism and Communism. It enshrines heterosexual marriage as normative and expressly declares that the state encourages childbearing. In addition, Article II states not only that "human dignity shall be inviolable" and that "every human being shall have the right to life and dignity" but also that "embryonic and foetal life shall be subject to protection from the moment of conception." Article III adds that "all practices aimed at eugenics . . . shall be prohibited."[63] The promulgation of the Constitution has not changed Hungarian law permitting abortion.[64] But it provides the possibility that it could—and certainly it is understood that way by both proponents and opponents of legal abortion access.[65]

In Poland in 2013 an effort to get rid of the fetal anomaly indication for abortion was defeated—by just one vote in parliament.[66] Again in 2016 Polish lawmakers attempted to outlaw all abortions, but this time they were met by massive nationwide protests and were forced to withdraw the legislation. In the wake of the defeat of the initiative, Prime Minister Szydlo—in a telling indication of just how potent the prodisability argument has become internationally for antiabortion activists—announced that "by the end of the year, the government would prepare a national program to support families with disabled children and women who give birth even after their fetuses have been found to have genetic disorders." The news was widely covered in the international press, with the right-wing *Fox News* in the United States headlining its contribution "Anti-abortion Poland Offers Payments for Disabled Newborns" (although explaining that the initial plan was for a one-time payment of 4,000 zlotys [approximately $1,000]).[67]

Many critics of abortion on grounds of disability—including Down syndrome advocacy groups like Don't Screen Us Out in the UK—are undoubtedly sincere in their conviction that such abortions are profoundly immoral and unacceptable (see figure 5). But there is no question that the only so recently achieved—and in many locales still quite shaky—consensus that the rights of individuals with disabilities deserve passionate defense (not to mention substantive financial investment) is being instrumentalized by individuals and groups with broader sexually conservative agendas. And it remains an open question whether this instrumentalization can in turn be used by disability rights advocates to generate greater support postbirth and across the lifespan.

Notably, the newest configuration in what is politically sayable and defensible with regard to "eugenics" or "abortion" has consequences for people far beyond those directly affected in some way by disability. This becomes clearest when we look at another of the most significant current

Figure 5. Don't Screen Us Out (2017). From a campaign launched in 2016 asking UK parliamentarians to oppose the introduction of a new noninvasive prenatal test (NIPT). Don't Screen Us Out coupled the call to halt the implementation of screening with a call for introduction of reforms to support individuals with Down syndrome and their families. Screenshot reprinted by permission of Don't Screen Us Out.

trends among individuals and groups committed to a sexually conservative agenda: it increasingly involves transnational, even transcontinental, mobilization. The changes have been rapid. Indicatively, when the European Parliament met in Strasbourg in 2002 to vote on a measure that would have encouraged all member countries to legalize abortion, the measure passed only narrowly—with a vote of 280 in favor versus 240 against—but it did pass. Eleven years later, the Estrela Report ("Report on Sexual and Reproductive Health and Rights," presented by the Portuguese member of the European Parliament, Edite Estrela), which called on all European nations to support reproductive and sexual self-determination for their citizens, including by providing sex education, was rejected. (Instead, an alternative conservative proposal that gave more rights to determine sex-related law and policy on the basis of the principle of "subsidiarity" to individual nation-states—the concept better known in the US context as "states' rights"—was accepted by a vote of 334 to 327, with 35 abstentions.)[68] This turn of events between 2002 and 2013 was not only due to the incorporation, in the interim, of formerly Communist nations into the European Union. Instead, credit was due, very specifically (and this despite the "subsidiarity"-based claim advanced that individual nation-states should have more leeway in the development of policy), to a number of transnationally organized right-wing NGOs.[69] Among the most significant of these are a group registered in Belgium and calling itself European Dignity Watch and the Spain-based HazteOír (Make Your Voice Heard).[70] Both are hostile to LGBT advocacy in combination with their opposition to reproductive self-determination.[71] (For instance, Sophia Kuby, executive director of European Dignity Watch, has declared that "sexuality only in a certain order has its proper place, namely in the marriage of a man and woman, in which there is the possibility of passing on new life. Homosexuality is thus always in deficit."[72]) But the most effective countermobilization to the Estrela Report came from the movement One of Us, which defends the protection of every fertilized egg as though it were already a citizen.[73] Although supported by many Roman Catholic Church leaders, One of Us makes its arguments for the protection of the fertilized ovum in entirely secular terms. One of Us is an example—indeed, the single most successful example—of a new phenomenon called the European citizens' initiative, which is designed to encourage democratic grassroots participation.[74]

One of Us is, additionally, represented by Grégor Puppinck, a French lawyer and director of the European Centre for Law and Justice, a right-wing NGO based in Strasbourg but founded (in 1998) by the US-based lawyer Jay Sekulow, head of the American Center for Law and Justice (itself in turn founded by the US right-wing evangelical minister and media mogul Pat Robertson), and it is clear that there is plenty of transatlantic cooperation and sharing of ideas between the two institutions and their directors.[75] Puppinck sits on various Council of Europe committees involved with questions relating to reproductive and sexual politics. The European Centre for Law and Justice initiates and supplies briefs in a steady stream of lawsuits involving reproductive and sexual issues at both the national and supranational levels; it holds conferences on such matters as late-term abortions (under the rubric of "neonatal infanticide"), "transgender marriage" (the individual seeking to marry is described as having undergone surgery in order to "resemble" a woman), surrogacy (referred to as "maternity traffic"), and reproductive testing and technology (summarily criticized under the rubric of "eugenics"), as well as on "religious conscience" clauses.[76] In addition, Puppinck regularly writes commentaries on legal and policy developments in European nations. For example, in his commentary of 2011 on the new Hungarian Constitution, including its assertion that life begins at conception and is under the protection of the state from that point on, Puppinck reiterated what he saw as the Constitution's merits. Evident was his appreciation for demographic concerns, his invocation of "Christian values" and of general gender and sexual conservatism (including a side swipe at LGBT rights activism), and his affirmation that "eugenics" must be resisted.[77]

A year later, 2012, the European Centre for Law and Justice and Puppinck weighed in on cases from Italy and Latvia. The European Court of Human Rights had just ruled that the Italian law against the screening of embryos conceived through in vitro fertilization should be overturned, with the argument that it was "incoherent" to permit abortion on eugenic grounds while forbidding preimplantation genetic diagnosis (PGD). The case in question involved a couple who had chosen to abort a fetus carrying cystic fibrosis (a life-threatening genetic disorder) after they had been denied the right to have the embryo tested before implantation. The court had found—as an antiabortion website summarized—that "the 'wish' to have a healthy child 'constitutes an aspect of their private and family life and comes under the protection of Article 8' of the European Convention on Human Rights" and that among the rights guaranteed by the convention was "a 'right to give birth to a child who does not suffer from the disease they are carriers of.'" Puppinck, in his amicus brief, expressed deep concern, noting that a "wish" had with the ruling become a "right." This, he contended, represented "a conception of human rights as a projection of the individual will in the social order." And this was unacceptable. The court's ruling had created, he said, "a right not to transmit bad genes, a right to eugenics."[78] In another, far more painful case (an instance of the type often grouped under the—awfully named—rubric of "wrongful birth"), a mother in Latvia had contended that her rights had been violated "when she was not offered genetic screening to help her decide whether to abort her daughter," who had been born with Down syndrome. The child had been born in 2002, but the mother had sought damages from the doctor for having failed to provide prenatal testing that would have permitted her to decide whether or not to terminate the pregnancy (for lost income due to her inability to work, as she had to care for the child, and for assistance with the cost of care itself) and subsequently from the Latvian state for failing to prosecute the doctor. The mother already had an older son who suffered from schizophrenia. The case had gone to the European court, which decided against granting the mother damages but did find that there had been procedural problems in the handling of the case at the national level—and while it did not find a right to abortion, it did find a right to "information." As the court was deciding, the European Centre for Law and Justice had intervened, joining with—in Puppinck's words—"a large number of European Down's Syndrome organizations . . . in a coalition opposing the creation of [what the Centre for Law and Justice called] a 'human right to eugenics.'" Together "they published a declaration named 'Stop Eugenics Now' presented at a conference hosted at the Council of Europe in July 2012."[79] Most recently, in March 2016 One of Us held the Pan-European Forum in Paris, which twelve hundred participants from twenty-eight European nations attended. At this event Thai surrogate mother Pattaramon Chanbua, who had kept the Down syndrome son, Gammy, whom she had borne for an Australian couple, was given the ONE OF US Award as a "humble hero of life."[80]

✐✐✐

The world of "reproductive torts" and "gen-etiquette" is changing daily.[81] But it is clear already that the European Convention on Human Rights' provisions, which do still include the right to privacy, the right to be self-determining in forming a family, and the right to dignity, understood as including a "protection from the use of women's bodies" for purposes against their will, are under growing pressure.[82] In these circumstances, a handful of commentators in various nations have sought to engender greater capacity for empathic identification with potential parents as well.[83] Some are lawyers, others bioethicists, philosophers, politicians, historians, and theologians—and still others counselors and affected parents, and affected individuals themselves. Some use reason; others appeal directly to emotions. All of them work to articulate a right to termination on grounds of disability that need not be interpreted as insensitive to disability rights (see figure 6).[84]

One such individual is a woman named Jane Fisher, who works for Antenatal Results and Choices (ARC) in the UK. This organization runs a telephone hotline for parents who have received a diagnosis of a fetal anomaly, and it provides nondirective, supportive counseling to help parents navigate the ensuing decision-making process about whether or not to terminate. Called to testify before the UK Parliamentary Inquiry on Abortion and Disability, Fisher reiterated again and again that choosing to terminate a pregnancy on grounds of a fetal anomaly like Down syndrome, with its broad and unpredictable spectrum of possible outcomes, or on grounds of a late-in-pregnancy discovery of severe brain damage by no means signaled any disrespect to living individuals with disability.[85]

As is becoming gradually more evident, what has in the current climate become most difficult but most necessary to say is that the demand that affected women or parents must not terminate and instead must embrace disability parenthood is morally presumptuous and unattuned not only to the responsibility that comes with giving life and the highly variable range of outcomes of any given disability but also to the vulnerability and precariousness of parents, whether they are potential carriers of predictable genetic vulnerability or they have been exposed to environmental toxicity or to a disease like Zika—not to mention parents experiencing socioeconomic or emotional precarity or who are confronted with a randomly appearing anomaly.[86] Across Europe—and, increasingly, also in the United States and worldwide—it has become

About ARC

- ▶ History
- ▶ ARC news
- ▶ Mission and vision
- ▶ Media area ▾
- ▶ Our sponsors
- ▶ Celebrity support
- ▶ Decision making
- ▶ Continuing a pregnancy ▾
- ▶ **Ending a pregnancy ▾**
  - ▶ What parents say about ending a pregnancy
- ▶ Another pregnancy
- ▶ Counselling
- ▶ Family and friends
- ▶ ARC forum
- ▶ Publications ▾
- ▶ In memory ▾
- ▶ Links

YOU ARE HERE: HOME > ABOUT ARC

About ARC, Antenatal Results & Choices

ARC is the only UK based charity helping parents and professionals through antenatal screening and its consequences.

OFFERS non-directive individualised information and support to parents who are making decisions around antenatal testing.

HELPS parents cope with the uncertainty and anxiety which is an inevitable part of the testing process.

What parents say about ending a pregnancy

Ending a pregnancy because of fetal anomaly can be a very isolating experience. The associated feelings can be complicated and sometimes overwhelming. It can be helpful to know your feelings are normal after such a traumatic experience. You may recognise some of the feelings described by these bereaved parents (all names have been changed):

Ann

'I will never forget the complete heartbreak of it all. We were faced not only with the fact that our baby was going to die, but also that we had to play a part in deciding the timing of his death.'

Lina

'It was agony, but we decided we would have to let our baby go as we are both in our forties and not having any family around us we felt it would be just too much for us to bring up a child with serious disability - and we had our other children to think of. We worried about who would look after our child after we were gone.'

Jo

'It was the most horrendous thing we have ever been through. The level of grief and emotional pain was frightening. But even after all this, ultimately we were grateful that we had the choice to end the pregnancy. We felt and still feel that we made the right decision for us, but also, importantly, for her.'

Jackie

'I have to say that the staff at our hospital were great. We were treated with sensitivity and sympathy throughout. But when I got home - with no baby - I was hit by a huge wave of grief. It really floored me and I know my partner was really worried about me. The ARC helpline was a lifeline then. I don't know how they put up with my endless sobbing, but they did.'

Clare

'After my termination for medical reasons last year I turned to ARC and was welcomed with open arms. At a time of despair you brought understanding, comfort and hope. You never judged me and made me realise I was not alone. When I reached those dark places you were the only ones I could count on.'

ever more difficult to celebrate diversity (including neurodiversity) and passionately to defend disability rights and adequate government financial support for dignified and flourishing lives for the physically, cognitively, and emotionally disabled and simultaneously to argue for the rights of pregnant women and their partners to choose whether or not to carry a disabled fetus to term. Yet, especially with regard to dignified and flourishing lives for adults with cognitive disabilities, in no nation are funding and supports remotely adequate.[87] Love and money both are all too often in short supply. This is a most pressing moral challenge for the present and future.

One puzzle is how this predicament and the perplexing constellation of alliances surrounding it first emerged and by what circuitous routes and unexpected intersections of historical trends particular interpretations of the lessons of the past ascended and have thereby now come to trouble the present. To begin to make sense of those issues, the next chapter turns to the specifically post-Nazi German dimensions of the present impasse. The final chapter expands again geographically to explore a variety of imaginative and inspirational forms of activism concretely advancing disability rights across the European Union.

Figure 6 (*left*). "What Parents Say about Ending a Pregnancy" (2017). Antenatal Results and Choices (ARC), UK, runs a telephone helpline and provides publications that offer nondirective information to parents before, during, and after antenatal screening. Another page of the website contains testimonials from parents "about continuing a pregnancy," and ARC also provides a safe-space forum for bereaved parents. Screenshot reprinted by permission of Antenatal Results and Choices (ARC), UK.

2

Moral Reasoning in the Wake
of Mass Murder

The Singer Affair and Reproductive Rights
in Germany, 1980s–1990s

We cripples will not let ourselves be used as propaganda objects! . . . The abortion opponents chose this spot in order to defame and to criminalize women who undergo an abortion, by putting the mass extermination of the disabled on the same level as abortion today. . . . This shameless equation is the reason we are appearing here and condemning the trivialization of the inhuman National Socialist murders.

**members of the Federal Association of Disabled and
Crippled Initiatives in a counterdemonstration at an antiabortion
rally at the former killing center of Hadamar, 1986**

In point of fact, a systematic and "legalized" murdering of the disabled has existed only under the Nazis, and they, simultaneously, punished abortion severely. In countries in which abortion has been liberalized, the disabled and the elderly and marginalized groups are generally treated with respect. There is no indication that abortion has or ever had anything to do with the killing of human beings.

reproductive rights activist Susanne von Paczensky, 1989

Kein Forum für den »Euthanasie«-Befürworter Peter Singer!

Am 26.5.2015 soll der australische Bioethiker Peter Singer in der Berliner URANIA einen Preis für seinen Beitrag zur »Tierleidminderung« erhalten. Bekannt ist Singer wegen seiner Forderung, die Tötung behinderter Säuglinge unter bestimmten Bedingungen zu legalisieren. Im April 2015 forderte er in einem US-Radio-Interview, behinderten Säuglingen Leistungen des öffentlichen Gesundheitssystems zu entziehen. »Ich möchte nicht, dass sich meine Versicherungsbeiträge erhöhen, damit Kinder ohne Aussicht auf Lebensqualität teure Behandlungen bekommen«, so der Princeton-Professor.

Leute, die für »Euthanasie« eintreten, sollten ihre Ideen nicht öffentlich verbreiten können.

›› Kein Preis und kein Forum für Peter Singer!

›› Gegen »Euthanasie« und Behindertenfeindlichkeit!

›› Keine Zusammenarbeit mit der Giordano-Bruno-Stiftung und ihren Ablegern!

Kommt alle zur Kundgebung:

Dienstag, den 26.5.2015
Ab 17.00 Uhr - Kleiststrasse / An der Urania

Kontakt: antisinger@riseup.net
Mehr Infos: no218nofundis.wordpress.com

V.i.S.d.P.: Petra Schreier, Kleiststraße 89 10787 Berlin

Figure 7. Poster protesting the awarding of a prize to philosopher Peter Singer in Germany, 2015. Reprinted by permission of the Aktionsbündnis "Kein Forum für Peter Singer."

In this chapter I turn to the case study of post-Nazi Germany at the pivotal moment near the end of the Cold War. While in the 1960s–1970s a negative attitude toward disability had facilitated the advance of abortion rights, for a complex conjunction of reasons the demand for a more positive attitude toward disability in the 1980s–1990s ended up having negative implications for abortion access. Almost as soon as abortion had been partially decriminalized across much of the Western world in the course of the 1970s, a backlash began to develop, slowly at first but then with gathering momentum. And while initially the activists driving that backlash had made no mention whatsoever of disability, a major national controversy over the lessons of the Nazi past caused a consequential reconfiguration in the terms of discussion over reproductive rights to be consolidated. Here again, but differently from before, we find an intricate and dynamic interplay of reverberations between past and present moments—and see as well how, in conflicts that are at once ideological and emotional, *interpretations* matter as much as, or more than, facts (see figure 7).

The Singer Affair

"Thou shalt not kill. That is not divine law, that's a Jewish invention."[1] So said Eugen Stähle, a medical doctor and head of the Division of Health within the Württemberg Ministry of the Interior when confronted in 1940 by Protestant religious leaders' protests against the first phase of the mass murders of the disabled that he was cocoordinating at that very moment. In this first phase, the meant-to-be-secret but by that point no-longer-so-secret program called by the Nazi leadership Aktion Gnadentod (Action Merciful Death) and generally, since the defeat of the Nazis in 1945, called Aktion T4 (in reference to Tiergartenstrasse 4 in Berlin, the address at which this program was planned), 70,273 individuals with psychiatric illnesses or cognitive deficiencies were, between January 1940 and August 1941, murdered with carbon monoxide in six specially designed gas chambers within what had been, previously, with the exception of one of the buildings, facilities for healing and care (see figure 8).

Figure 8. Hadamar in 1940–41 with the crematorium chimney smoking. The town of Hadamar is about an hour's drive from Frankfurt am Main; the elegant building on a hilltop had formerly been a psychiatric hospital. Through decades of intensive postwar activism, the name Hadamar has come symbolically to represent the mass murders of individuals with disabilities or psychiatric illnesses in a way comparable to the representative role of the name Auschwitz for the Holocaust of European Jewry. Reprinted by permission of the Diocesan Archive Limburg, Nachlass Pfr. Hans Becker; photograph: Wilhelm Reusch.

Eventually, due to unrest in the populace and further religious protest—especially the prominent Catholic bishop Clemens August von Galen's sermon of August 1941 decrying the killings—Hitler had ordered the program officially stopped. It continued on nonetheless in a second, decentralized phase that lasted even beyond the end of the war in May 1945.[2] Ultimately, 210,000 individuals with intellectual or psychological disabilities in the German Reich and a further 80,000 in occupied Poland and the Soviet Union were killed through deliberate medication overdose, poisoning, or systematic starvation.[3] Meanwhile, the approximately one hundred personnel who had gotten their training and practice in murdering the disabled in the T4 facilities, along with their now field-tested technology of carbon monoxide gas chambers, were moved on to Poland to turn their attention to the mass murder of European Jewry in the Operation Reinhard death factories of Belzec, Sobibor, and Treblinka (where approximately one-quarter of the Holocaust took place) (see figures 9a and b).[4]

The Stähle quote—"Thou shalt not kill. That is not divine law, that's a Jewish invention" (in other words, Moses just fabricated the Ten Commandments, and no self-respecting Nazi need concern himself with them)—was brought to the attention of West German readers of the weekly *Die Zeit* in the summer of 1989 by Ernst Klee, a prominent nonacademic historian and advocate for the rights of the disabled in the context of Klee's vehement and eloquent repudiation of the theories of the Australian philosopher and animal rights activist Peter Singer. Singer had been invited to West Germany by an organization called Lebenshilfe (Life Assistance), the premier association of parents and caregivers of disabled children, in the expectation that he would address a scheduled conference in Marburg titled "Biotechnology—Ethics—Mental Disability"; he had in addition been invited by the special education expert Christoph Anstötz, a professor at the university in Dortmund, to speak there on the subject "Do severely disabled newborn infants have a right to life?" (Singer's own short answer to this question was no, as the second sentence of his then recently published book, *Should the Baby Live?* [1985], cowritten with the philosopher Helga Kuhse, also Australian, though of German heritage, stated clearly: "We think that some infants with severe disabilities should be killed."[5]) It was not, the organizers later said, the outpouring of indignant letters from across the land but rather the announced threat that there would be demonstrations and public disruptions of the proceedings that caused both invitations to Singer to be withdrawn.[6]

Only in one German town, Saarbrücken, was Singer able to participate in a public discussion with his local hosts, the philosophers Georg Meggle and Christoph Fehige. (That event also began with a half hour of ear-piercing whistles and shouts demanding "Fascist out!") There in Saarbrücken, the audience was able to learn, among other things, that

Figures 9a and b. The train station at Treblinka (*top*) and rail personnel at Sobibor (*bottom*). Altogether, 1.5 million Jews were murdered with carbon monoxide generated by vehicle engines at the Operation Reinhard camps of Belzec, Treblinka, and Sobibor. Reprinted by permission of Yad Vashem Photo Archive and Photo Archives at the United States Holocaust Memorial Museum.

Singer was the son of Jewish refugees from Vienna, that three of his grandparents had been murdered in Nazi concentration camps, and that he resolutely defended his conviction that since passive killing of severely disabled newborns by withholding treatment was already quietly being practiced in hospitals across Germany and elsewhere in the Western world, active mercy killing by doctors to shorten these newborns' agony should be permitted as well, within strict limits.[7] Moreover, Singer noted that he was of the opinion that certainly conditions for already-living individuals with disabilities should be better. ("Often the conditions in homes for disabled people are terrible. As a prosperous society we should do more to improve the quality of life of disabled people, and we should also try as much as possible to integrate them into society."[8]) Singer's critics, however, were far from mollified.

Because massive media coverage had accompanied the controversy from the start, with *Die Zeit*, for instance, titling one early contribution "Can Euthanasia Be Defended on Ethical Grounds?" (indeed, the paper's own tilt toward answering yes, although quite explicable within its own terms, was part of what had caused Klee to write his countervailing piece), over the following months and into the next year the ramifications kept expanding as local and regional papers published articles with titles such as "Parallels to Nazi Theories," "Fury and Outrage at the University: Protest against 'Academic Chairs for Euthanasia,'" and "We Are Afraid for Our Children."[9] The Green Party issued a statement referring to Singer's theories as an "incitement to murder."[10] Foreign observers from the United States and the UK expressed shock that civil conversation about ideas was, apparently, impossible.[11] The whole thing, depending on how you looked at it, turned out to be a fiasco for the would-be hosts or, as Singer's defenders argued, a sign that a tiny minority of overreactive extremists—who had not even read Singer closely—could shut down rational debate for an entire country. Singer himself, piqued, wrote a piece in the *New York Review of Books*, "On Being Silenced in Germany."[12] And Anstötz, who had originally hoped to host Singer in Dortmund (but instead had faced livid protesters—from religious representatives to the main AIDS organization—with banners declaring "Boycott Anstötz's Murder Seminar," "No Murder of Babies, the Elderly and

Disabled," and "For Anstötz and Singer, Disabled Newborns Are Human Vegetables"), subsequently copublished a collection of documents about the confrontations—*Peter Singer in Deutschland*—whose subtitle, *The Endangerment of Freedom of Discussion in Scholarship*, announced its concerns.[13]

Over and over, the very fact that there had been mass murder of people with disabilities in the nation's past was put forward by Singer's defenders as a main explanation for what was asserted to be the immaturity of moral reasoning abilities in West German society in comparison with the rest of the West, a lamentable and inappropriate oversensitivity that led to "thought and discussion taboo[s]," an incapacity to confront the genuine and inescapable challenges brought by technological advances and crises of extremity of suffering at either end of life.[14] For Anstötz, moreover, it was the *critics'* refusal to let Singer speak that was best compared to the Nazis' "burning of books."[15] He and his coauthors contended that the very characteristics Nazis had ascribed to Jews ("sly outfoxing reasoning," "analyzing, distanced, holds nothing sacred . . . emphasizes logic")—characteristics that the Nazis had been determined to "exterminate" (*ausrotten*)—were still, sadly, lacking decades later ("even a superficial glance at the political culture of Germany shows how thoroughly, and with legacies continuing into the present, this extermination has succeeded"). On this basis, in turn, they concluded: "Conversely, it becomes clear how urgently we need precisely that spirit of reflection, of clarity, of analysis, of differentiation and of tolerance that is embodied by Peter Singer."[16] And Georg Meggle, who had hosted Singer in Saarbrücken, wrote that Singer's critics were promulgating "a new form of antisemitism," charging the critics with assuming that "if a Jew thinks like Singer thinks, then he must be sick."[17] Singer, too, weighed in, writing in *Bioethics* in 1990 that "perhaps what really was instrumental in preparing the Nazi path to genocide, and has not yet been eradicated in the modern Germany, is not the euthanasia movement at all, but the kind of fanatical certainty in one's own rectitude that refuses to listen to, or engage in rational debate with, anyone who harbours contrary views."[18] Nonetheless, the critics would have the last word. As a radical disability rights newspaper, *Die Randschau*, declared also in 1990: "The 'tolerance for debate' that the philosophers are demanding for Singer's theses is the same as one that would permit the discussion of the thesis of the 'superiority of the Aryan race.' But in both cases, the fundamental will to treat human beings as unequal must be combatted."[19] This was to remain the general tenor of what would become a broadly propounded official anti-Singer stance. It would be a full fifteen years before Singer delivered another lecture in Germany.

Post-Nazi Politics and Historiographical Frames

I began with Klee's and others' stinging rebukes to Singer—or rather, with what came to be called "the Singer affair"—for multiple reasons.[20] One reason is that the Nazi doctor Stähle's quote, which the historian Klee had uncovered as he was researching the Nazi murders of the disabled for his magnum opus on the topic (in English the title would be *"Euthanasia" in the National Socialist State: The "Extermination of Life Unworthy of Life"* [1983]), captures with unintentional transparency the intimate interconnections between antisemitism, on the one hand, and contempt for individuals with disabilities, on the other.[21] One of the great and consequential dramas of the 1980s and 1990s, in scholarship internationally and in activism within Germany, would be the determined effort to elucidate the multiple links—in staffing, in gassing technology, but also in attitude toward "lives unworthy of life"—between the murder of individuals with disabilities and the Holocaust of European Jewry. Indeed the quote and Klee's use of it bring into view just how very important the invocation of these two interrelated (or, as historian and survivor of Auschwitz Henry Friedlander put it, "intradependent") mass murders in the nation's past would be for advancing the cause of disability rights in the postwar then-present of the 1980s.[22] It is hard to remember now but crucial to our understanding of the dynamics at the time that contempt for and cruelty toward the physically and cognitively disabled lasted well into the 1980s (and even beyond). The postwar years had seen a (in hindsight truly stunning, then simply devastating) breadth of popular support for the perpetrators and ongoing shaming of the victims and their families.[23] Few of the perpetrators ever faced justice; instead; they had illustrious postwar careers.[24] The very statements I asserted as facts in the opening paragraphs of this chapter—that the mass murder of the disabled was the precursor to and continued to be entangled with the Holocaust—were not generally obvious in the 1980s. Indeed initially, connections had been made more by intuitive emotional analogy than by specifying literal links (see figure 10).[25] This was a connection that still needed to be solidified and concretized in the public mind; the Singer affair provided a major occasion for doing so.[26]

Singer himself, in his widely used textbook of 1979, *Practical Ethics* (translated into German in 1984), had argued strenuously—and in this he was in accord with the assumptions animating much late 1970s scholarship—that there was no connection: "If euthanasia somehow leads to the Nazi atrocities that would be a reason for condemning euthanasia. But is euthanasia—rather than, for example, racism—to be blamed for the mass murders the Nazis carried out?" Singer's own answer to the question as he framed it was no. For him, hostile or lethal treatment of the disabled simply did not count as racism.[27] It was precisely this presumption of a categorical gulf between the two major Nazi murder programs that, it was felt, needed to be challenged, and it was over the course of the 1980s, through sustained research and advocacy work, that the links were starting to be established. Increasingly, moreover, a second tie was forged: conceptual and empirical connections were elaborated between the four hundred thousand coercive "eugenic" sterilizations of individuals with disabilities enacted under the rubric of the July 1933 Law for the Prevention of Hereditarily Diseased Offspring and the two-hundred-thousand-plus "euthanasia" murders.[28]

As it happened, the postwar government had continuously refused to acknowledge the harm done to victims of coercive sterilizations, rejecting their claims to be "persecutees of the Nazi regime" deserving of any recognition, much less of financial recompense, and relying on the opinion of experts, some of them ex-perpetrators (including those involved in the euthanasia murders), in declaring the sterilization legislation to have nothing to do with "National Socialist racial laws."[29] But Protestant Church leaders had not offered a countervailing moral position. Instead, eager to advance their own version of a sexually conservative

Figure 10. Demonstrator in a wheelchair wearing a yellow star with a wheelchair symbol inside at the first nationwide West German disability rights demonstration in Frankfurt am Main, 1980. As Carol Poore has explained in *Disability in Twentieth-Century German Culture*, the demonstration was held in outraged reaction to the discriminatory verdict of a court that had granted a nondisabled woman a refund of half of her expenses for a

"personal eugenics" in the postwar years and instrumentally invoking their unabashed pride in having resisted, however ineffectually, the murders in order to advance their own advocacy for "voluntary" sterilizations, they worked hard—and successfully—to keep "eugenics" and "euthanasia" analytically distinct.[30] It was *against* these trends of the first three postwar decades that a historiography arose, over the course of the 1980s, that reframed the Third Reich in such a way that both eugenics and euthanasia would come to be seen as central rather than marginal aspects of what was finally, by 1991, shorthanded (in historians Michael Burleigh and Wolfgang Wippermann's title) as "the racial state."[31] Burleigh and Wippermann, building on a decade of pioneering scholarship, expressly identified the Nazi goal as "the 'purification of the body of the nation' from 'alien,' 'hereditarily ill,' or 'asocial' 'elements'" and thus focused their account on "all those whose lives or reproductive capacity were ended as a result of Nazi racial policy," including "Jews, Sinti and Roma, and members of other ethnic minorities categorized as 'alien,' as well as the 'hereditarily ill,' 'community aliens,' and homosexuals." Indeed, they said, "there is much evidence to suggest that race was meant to supplant class as the primary organizing principle in society."[32] The debates around Singer had finally made this kind of summary statement seem incontrovertible, as major news outlets had taken the critics' cues and had begun in 1989, in text and in accompanying imagery, to center the murder of the disabled at the heart of the Third Reich and to register that "eugenic thinking" needed, on a regular basis, at least formally to be repudiated as immoral.[33] (Only since the turn of the millennium has the newest research led to the prospect of once more decoupling eugenics from euthanasia and to the prospect of debiologizing the Third Reich more generally.[34])

—————

vacation she had taken in Greece at which—she alleged—she had been forced to see individuals with physical and cognitive disabilities staying at the same hotel. The court's decision had read in part: "It is undeniable that the presence of a group of the severely disabled can reduce the enjoyment of a vacation for sensitive people." Demonstrators also carried a banner declaring: "Don't pity the disabled person; pity the society that rejects him." Photo © 1980, 2017 by Walter Pehle.

My second purpose, however, in revisiting the fallout from the Singer affair and situating it in its various overlapping contexts is that doing so helps us understand not only the particular shape taken by radical disability rights activism in West Germany in the 1980s–1990s and the particular ardent investments the movement developed but also their complicatedly ricocheting consequences. For it was, of all people, Singer, whose convoluted mix of mundanely sensible and traumatizingly obscene lines of moral reasoning, coming at the historical juncture that he did, *who created the opportunity for radical disability activism in West Germany to erupt into mainstream public view*, garnering the attention — and respect — of both major media outlets and government officials and becoming a political force to be reckoned with.[35] But no less significant is the impact of the debates about Singer's theses *on the terms in which women's rights to access abortion could be defended* — rights that were, co-incidentally, at that very moment in 1989 under renewed attack from conservative forces and about to be yet more fully reconfigured after the collapse of Communism just a few months later. It would become impossible, in the wake of the Singer affair, for any mainstream German politician frankly to defend abortion on grounds of what had been called the "eugenic" or "embryopathic" indication; by the early 1990s politicians had backed away from any language that might possibly be construed as suggesting a diminished respect for disabled life — and so too had many feminists. There was a rush to outdo one another in declaring that the state should not and would not ever prefer nondisabled over disabled life. Disability activists were key players in the reorientation. This particular fallout was not inevitable, but it was overdetermined.

Different outcomes might have been possible. In other national contexts, including in the United States, the UK, Austria, France, Switzerland, and Israel, it has been imaginable that one could passionately defend disabled newborns' and disabled children's and disabled adults' rights to life — indeed joyful, rich, beloved, and fulfilled life — and simultaneously find completely morally acceptable abortion on grounds of fetal disability. In Israel this is deemed "the two-fold view of disability."[36] But something — many things — about the particular conjunction of circumstances in Germany at the transition from the end of the

Cold War to the new unified nation made this perspective seem, quite apparently, unfathomable. Feminist efforts to mobilize against the strong rollback under way against the partial decriminalization of abortion that had been achieved in 1976 collided head-on into the debacle around Singer.

What was it that had so alarmed the protesters against Singer? Initially, the mainstream media had been nonplussed at the uproar. His *Practical Ethics* seemed pertinent to deliberations that had been already ongoing in West Germany for the prior ten to fifteen years involving dilemmas surrounding technological advances in end-of-life care, as well as patient requests for assisted suicide.[37] His suggestion that permitting doctors to provide active killing rather than extending a severely disabled newborn's torment through passive letting-die (for example, in cases of inoperable spina bifida), though instinctively repellent to and immediately repudiated by many, at least seemed discussable.[38] And finally—though it took a while for the major newsmagazines and newspapers to make much of this—Singer was a staunch advocate for animal rights, and although the vast majority of West Germans were meat eaters, there were also numerous dog lovers, and there were certainly broad sectors of the populace that would be receptive to, or at least not agitated about, arguments for humane treatment of animals.

But the trouble lay in the way Singer *joined* his various areas of interest. Singer could easily have argued that animals—nonhuman sentient beings—deserved far better treatment than humans normally meted out to them and left it at that. Instead, Singer repeatedly evinced (indeed, nearly obsessively reiterated) a preoccupation with denigrating the cognitively disabled, stating that severely cognitively disabled individuals lacked "personhood" and hence had *less* value and less right to life than animals (many of whom he thought *did* have the "personhood" the cognitively disabled were missing). Thus, for instance, in an essay from 1983 titled "Sanctity of Life or Quality of Life?" Singer had stated: "If we compare a severely disabled human child with a nonhuman animal, for example a dog or a pig, we will frequently find that the animal demonstrates higher capacities with respect to comprehension, self-consciousness, communication and many other things."[39] And in

Practical Ethics Singer had argued that even if someone belonged to the human species, he or she was "not a person," if "rationality, autonomy and self-awareness" were absent. Or connecting the dots more explicitly: "Some members of other species are persons: some members of our own species are not. No objective assessment can give greater value to the lives of members of our species who are not persons than to the lives of members of other species who are. On the contrary, as we have seen there are strong arguments for placing the lives of persons above the lives of nonpersons. So it seems that killing, say, a chimpanzee is worse than the killing of a gravely defective human who is not a person."[40] These were the kinds of comments that made his critics apoplectic.

Klee, in his brilliant rejoinder to Singer, homed right in on what he called the "bizarre nexus of animal rights and euthanasia" in Singer's work and regaled his readers with examples of *Nazis* who had linked their enthusiastic embrace of animal rights both with antisemitism (Hitler was the "savior" of animals from "Jewish materialistic" "animal torture" like vivisection) and with lethal antidisability sentiment (the SS journal *Das schwarze Korps*, for example, had declared with regard to "mercy killing": "A child born as an idiot has no value as a person. . . . He is less aware of his existence than an animal").[41] Franz Christoph, one of the cofounders of the radical "cripple movement" (*Krüppelbewegung*), launched in the 1970s, in his own extended rebuttal to Singer in the pages of the newsmagazine *Der Spiegel* also made the comparison to Nazism. (Christoph, a polio survivor, had already made a name for himself in 1981, when he had the audacity to strike the federal president, Karl Carstens, a former Nazi, with a crutch at the occasion of paternalistic government festivities in Düsseldorf organized in keeping with the UN declaration that 1981 should be the Year of the Disabled, and he had a multitude of creative protest actions to his credit.)[42] Invoking Singer's opinion that "the killing of a disabled infant is not morally equivalent to the killing of a person. Very often it is no injustice at all," Christoph observed curtly:

> In connection with any other group of people the thesis of Singer would be in danger of being rejected as fascistic thinking—without

any scholarly dialogue. An event that wanted to concern itself with the right to life of newborn women or foreigners would—justly—bring in its train a storm of public protest and would, if it were to be debated at a university, have to reckon with the intervention of the relevant government minister. Also the minister of the family in Bonn would be careful not to sponsor such an event—if it was not a matter of the cognitively disabled.[43]

For Christoph, the most urgent task was to respond to defenders of Singer who liked to gesture (vaguely, and impatiently) to the Nazi past as the reason for the (in their view regrettable) touchiness and reluctance to discuss Singer's theses, and he targeted (as cynical) Singer's own pronouncement to the effect that "we cannot condemn euthanasia just because the Nazis did it, any more than we can condemn the building of new roads for this reason." Christoph was intent on putting forward a *different* interpretation of how the Nazi past mattered—not, as Singer's proponents claimed, because it caused German conversations to be out of step with international trends around assisted suicide and related matters but rather to articulate why *talk* could be so offensive. It was, Christoph said, "precisely these kinds of scholarly discussions and discourses that were precursors of what came to be, beginning fifty years ago, the extermination of 'life unworthy of life.'" The trouble was the way that a question was being established as even legitimately posable, the very act of asking "'Euthanasia for severely disabled newborns?' that then could be answered with a yes as well as with a no." As one woman in the cripple movement had phrased the point succinctly, as quoted by Christoph: "We cannot just tolerate it, when they talk about whether we may live or not. . . . It is an incredible intolerance, when the human dignity and personhood of disabled people is massively assaulted, when the violability of human life is therewith even more socially legitimated." Christoph's conclusion was thus that it was "specifically because of the historical experience, although social service bureaucrats apparently cannot relate to this," that "for those who are affected, any and all discourse about the reintroduction of the concept of 'life unworthy of life' seems like a menace to their right to live." *Der Spiegel* took the cue and accompanied Christoph's piece not only with a photograph of a sit-in to disrupt a rehab experts' conference on the topic of assisted suicide in Karlsruhe the year before (where Christoph and others had worn blue garbage bags, and Christoph had worn a sign around his neck declaring, "I am unworthy of life") but also with a photograph of the distinctive gray buses that had brought the disabled to their deaths in the Nazis' T4 program and with a copy of Hitler's order, backdated to the start of the war on Poland in September 1939, that permitted the beginning of the calculated murder of five thousand disabled children.[44] (Singer later took particular umbrage at the magazine's decision to use these supplemental images.[45]) The tide of mainstream consensus was suddenly but manifestly turning—after an excruciatingly long delay of four postwar decades—in favor of radical disability activists' views on the proper lessons to be drawn from the Nazi past.

Abortion versus Infanticide

In addition, however—and although the broader potential implications of what had initially seemed like a side note in Singer would not become apparent until several years later—Singer had used as a springboard for his own causes something that had actually been an achievement of feminist and sex rights activism just a few years before his book was published. It was specifically the fact that across the Western world, abortion had—due to vigorous women's rights advocacy—been at least partially decriminalized and had come to be seen as morally acceptable by broad popular majorities that Singer used as his entry point for theorizing the acceptability of active infanticide (again, with frequent interpositions making comparisons with animals). Over and over, he had made a case for seeing the similarity, rather than the difference, between "killing the late fetus" and "killing the newborn infant." Thus, for instance—in this moment speaking about all newborns, not just disabled ones—Singer expressly built his argument on the basis of the only just recently established greater moral acceptability of abortion. "If the fetus does not have the same claim to life as a person, it appears that the newborn baby does not either," he began one sentence, going on from there to assert once more that "the life of a newborn baby is of less value than the life of a pig, a dog, or a chimpanzee." But the difference *between* newborns had to do with parents' desires for them, and Singer assumed that parents desired the disabled less. Moreover, then, in Singer's view, and since not all disabilities were evident prenatally—some indeed might be caused in the birth process—in cases of disability (and he was ambiguous about what counted as severe) sometimes active infanticide should be permitted, perhaps up to "a month" after birth. And at yet another moment, and while Singer would, in the ensuing controversies in Germany, keep insisting that he had *never* argued for the killing of already-living disabled individuals older than infants, readers could be forgiven for thinking that he actually had. "For simplicity," he had written in *Practical Ethics*, "I shall concentrate on infants, although everything I say about them would apply to older children or adults whose mental age remains that of an infant."[46]

This insistence on not drawing the line either at birth or at viability but instead actively blurring the boundary between abortion and infanticide was to have tremendous consequences not just for the reception of Singer in that summer of 1989 but for the reconfiguration of women's access to abortion that was, after the collapse of Communism just a few months later, shortly to ensue. In the complex back-and-forth between constituencies that followed, feminists would lose the ability to retain the—morally crucial—distinction between an abortion on grounds of anticipated disability and an infanticide. As repelled as most activists on behalf of both disability rights and women's rights were by Singer, many appear to have accepted his terms of debate.[47]

The antiabortion movement in West Germany had been trying since at least the early 1980s to involve disability rights groups in their political agenda, not only floating once more the time-honored (truly, since 1946!) maxim that abortions were somehow comparable to Auschwitz but also trying to make the link between abortions and the murder of the disabled—focusing specifically on the "eugenic indication" for abortion as a reason for disability rights activists to join them and speaking of the "thousandfold killing of unborn disabled babies" and of how "the so-called amniocentesis provides the ammunition for the fatal shot."[48] Among other things, a group calling itself Movement for Life had managed to organize individuals with disabilities in an affiliate called the Helen Keller Circle—and at least one young disabled man had written an open letter to the president of the Federal Republic criticizing the way "amniocentesis differentiates between 'worthy' and 'unworthy' life" and comparing the "eugenic indication" for abortion to Hitler's 1939 directive to begin the euthanasia killings.[49] Initially, radical disability groups spurned these overtures. For instance, in 1981, when prompted by a Heidelberg-based Catholic antiabortion student association calling itself Working Group for Life and condemning abortion under the slogan "Thou shalt not kill!" the Action Group against the UN Year of the Disabled ("against" because disgusted by what it took to be self-congratulatory but condescending and repressive charity efforts sponsored in that UN Year) responded:

> On the basis of our experiences as cripples and as nondisabled but concerned individuals in our society, we do not presume to condemn women who decide against a disabled child. Because of the hostility to disability in this society in particular, parents of disabled children are left alone with their problems and difficulties. Often the only recourse is to give the disabled child into an institution. If parents try to protect their child from an institution, this means taking up an immense battle with the authorities, social service offices, doctors and bureaucracy, battling against prejudice, financial challenges and isolation.[50]

In 1983 and again in 1985, when the antiabortion group Action for Life reached out to the Federal Association of Disabled and Crippled Initiatives "with an alliance proposition," the Cripple Group Bremen reacted negatively. Although the Bremen activists concurred that abortion on grounds of fetal disability was a problem for them ("we are . . . opposed to . . . the eugenic indication"), they rejected the campaign to criminalize all abortions: "To get rid of Paragraph 218 [the law on abortion] . . . would change nothing in the life reality of cripples in our society. We would continue to be disenfranchised and separated out, 'social euthanasia' would continue to occur. Thus we see our immediate task in improving the life conditions of those already living, and we would welcome it greatly if organizations like yours would also engage themselves in this direction." Moreover, "we have no desire to let ourselves be instrumentalized for your battle against ¶218." And in addition: "We find the comparison drawn . . . between ¶218 and Auschwitz conspicuously tasteless. Aside from the fact that 'euthanasia' did not take place in Auschwitz, we find this comparison to make a mockery of the victims and survivors of the concentration camps."[51]

In 1985, when feminist members of the "cripple movement" compiled a book of essays on the particular difficulties confronting women with disabilities, the book, *Geschlecht: Behindert, Besonderes Merkmal: Frau* (Gender: Disabled, special characteristic: Woman) included an illustration indicating how women in wheelchairs were working to abolish Paragraph 218 (see figure 11).[52] As late as 1986, when the antiabortion group Action for Life had decided not only to hold its annual demonstration at Hadamar—one of only two of the former six Nazi killing centers that were on West German soil—but also to invite disability activists to join them, the radical cripple activist Gisel Hermes published an incensed response in *Die Randschau*. Hermes explained to its readers the right-wing, gender-conservative, and antiforeigner racist values animating the hard core of the antiabortion movement and denounced the way "we so apparently are being used as show-pieces for an action that trivializes the fascist crimes against the disabled." Hermes was clear that "for the opponents of abortion, only the unborn, not the born life, appears worthy of protection."[53] On the day of the demonstration and counterdemonstration, members of the Federal Association of Disabled and Crippled Initiatives spoke out about how "the abortion opponents chose this spot in order to defame and to criminalize women who undergo an abortion by putting the mass extermination of the disabled on the same level as abortion today." They rejected "this shameless equation," pointing out further how the same conservative politicians who were working to erode abortion rights were also cutting funding for the very social services the living disabled so badly needed: "We cripples will not let ourselves be used as propaganda objects!"[54]

Yet that same year, 1986, other feminists within the radical cripple movement were already reporting on their dismay and anger at what they saw as too many nondisabled feminists' refusal to join in with a

Figure 11. Drawing of women in wheelchairs dismantling Paragraph 218—the paragraph that regulates abortion. Silke Boll et al., *Geschlecht: Behindert, besonderes Merkmal: Frau* (1985).

critique of the "eugenic indication." "A wall goes up," Swantje Köbsell and Monika Strahl explained, and "they block off and refuse to engage the actual problematic," "they accuse us of being opponents of abortion, . . . but we are not against abortion per se, only against the aborting, as a matter of course, of fetuses that have been declared as 'defective' and therefore undesired."[55] This was the compromise position that would come to define the feminist disability movement. Abortion on any grounds aside from anticipated disability would be adamantly defended;

abortion because of anticipated disability would be rigorously, righteously rejected.[56]

Other—both male and female—members of the cripple movement, as well as its nondisabled supporters, would often be yet harsher in the language they chose. Over and over, the idea of aborting on grounds of disability was presented as in and of itself hurtful to living disabled individuals.[57] Christoph had, for instance, written: "It makes me aggressive when I hear women say 'I have nothing against the disabled, I accept them completely, but if I were to be expecting a disabled child, it should, if possible, be aborted!' That reminds me of the father who has nothing against blacks but nonetheless throws his daughter out of the house if she is engaging in hanky-panky with a 'negro.'"[58] And in the wake of the Singer affair, in the pages of a journal on special education, the disability specialist Peter Rödler said that an abortion must be left up to the woman and her partner, but if the decision was made on grounds of knowledge of fetal disability, then "it is murder."[59] Both male and female activists called on women to boycott prenatal screenings as though to do so was in itself a moral imperative. In contrapuntal tandem with the antiabortion movement they otherwise despised, then, disability activists would come to develop—and would do so even more emphatically in the aftermath of the tumult over Singer—a *historically wholly new* singling-out for special condemnation of abortion on grounds of fetal abnormality, extending their repulsion at the proposal for active infanticide backward into the pregnancy.

Already before Singer hit the news, in a roundtable published in the New Left journal *Konkret* in April 1989, the well-known feminist journalist and cofounder of a family planning clinic in Hamburg, Susanne von Paczensky, discussed the new perceived impasse between disability rights and women's rights with Christoph and with three other women: the Green / Alternative List feminist Adrienne Goehler; feminist author Katja Lehrer, best known for a book on "raven mothers" (the German expression for women who refused to perform a traditional self-sacrificial maternal role); and Hannelore Witkofski, a member of the Disability Forum and another radical cripple organization, AG SPAK. Von Paczensky saw through the emergent conundrums and made the point that the very existence of the exceptional circumstance surrounding the "eugenic indication"—allowing abortions in cases of anticipated disability up to twenty-two weeks, rather than the twelve weeks for the other indications—had been a sign above all of lawmakers' concern not with whatever difficulties would ensue for an individual mother but rather with the concern that a disabled individual would be a burden on the national economy and thus for that reason his or her birth should be prevented. Indeed, she was convinced that the existence of the "eugenic indication" was in fact a sign of how "extremely hostile to the disabled" German law and culture were. But she also thought the animus against women who sought prenatal diagnostics—and then in "1–2 percent" of cases went on to choose abortion—was inappropriate and overwrought. She held fast to her conviction that whether a woman had an abortion because she did not like children, or did not want one at a particular moment in her life, or did not want a disabled child, "that is okay. . . . We are not authorized to judge on what grounds women abort." Witkofski, by contrast, openly charged that "cripples are being selected away before birth" and that this represented an attack "on my own life." When challenged, she said explicitly that yes, she was in favor of a "compulsory birthing of cripples." Goehler, in reaction, noted that she found it "unbelievably brutal" for someone like Witkofski to say that a woman who was deciding against carrying a pregnancy with a disabled fetus to term was somehow implying that she was "trying to get rid of all cripples in the world," as "though people also wanted, after the fact, to abort you": "We cannot solve the problem of a cripple-hostile society on the backs of individual women." Witkofski, however, was adamant. Women who aborted what she in furious sarcasm referred to as "that cripple stuff" were, in her view, "perpetrators." Only von Paczensky, who, incidentally, was half-Jewish (she survived the Third Reich not least because her non-Jewish mother had refused to divorce her Jewish father and thus had counted as a "mixed" individual, a *Mischling*, and thereby avoided deportation), pointed out that solely the Nazis had murdered disabled people—and simultaneously they had punished abortion severely. Moreover, she noted, "in countries in which abortion has been liberalized, the disabled and the elderly and marginalized groups are generally treated with respect." Her bottom line: "There is no indication that abortion has or ever had anything to do with the killing of human beings."[60] But this would prove to be a losing position.

In general, nondisabled feminists were ill equipped to respond to the conservative attacks on abortion once the flap over Singer had exacerbated the situation.[61] Increasingly, the New Left (or what was left of it in the "alternative scene," as well as the Green Party, its partial offshoot) was on the defensive for having displayed a hugely disability-insensitive preoccupation with "healthiness." As they had before, but especially in the wake of the nuclear reactor explosion at Chernobyl in 1986, New Left and feminist periodicals had published some extraordinarily offensive images that lampooned disability as a likely outcome of technology run amok. For example, *Konkret* had published a (montaged) photograph of a woman without arms mockingly captioned "My mother was for thalidomide. Now she's for atomic energy. Maybe this time she'll be right."[62] Meanwhile, although they had occasionally argued with disability rights activists over abortion rights, a majority of feminists of the era shared with disability rights activists a reflexive distrust of reproductive technologies—these (as they were called) "newest inventions of the techno-patriarchy."[63] Already by the fall/winter of 1988, when Green Party feminists held hearings about abortion in the Bundestag—urgently trying to collect arguments *against* the accusation that abortion was murder—they had also invited the disability activist Swantje Köbsell to address them and, moreover, concurred, as though it was self-evident, that reproductive technologies were profoundly immoral.[64] By January 1990 dozens of feminist antireproductive technology groups, anti-fascist collectives, and prominent post–New Left journalists—including the editor of *Konkret*—had signed a declaration published in the leftist Berlin daily, *die taz*, against Singer's right to speak.[65]

Changing the Law

But how did *lawmakers*—across almost the entire ideological spectrum— come to adopt the compromise formation first formulated by feminists in the cripple movement?[66] A complicated set of intersecting dynamics led to a fundamentally reshaped legal landscape between the fall of the Wall in November 1989 and October 1995, when a new law for a unified Germany was approved.[67] In the wake of the decision to unify in 1990, western feminists had hoped that the accession of the former eastern states, where abortion had been decriminalized in the first trimester since 1972, would lead to the adoption of a more liberalized handling of abortion also in the West; but conservative lawmakers—particularly in Bavaria, which was dominated by the Christian Social Union—had, that same year, preemptively sought the opinion of the Constitutional Court in anticipation of a possible attempted liberalization while simultaneously providing their own arguments against abortions on all but the most narrowly construed grounds.[68]

The Bundestag had indeed in 1992 promulgated a law that decriminalized an abortion when it could be shown "to prevent a danger to the life or physical or mental health of a pregnant woman," but the hopes of western and eastern feminists were dashed in 1993, when this law was voided by the Constitutional Court (relying in part on the Bavarian state's positions) on the argument that abortion *must* officially remain criminalized because of the Basic Law's guarantee of "protection of life" and that indeed women, in almost all circumstances, had "an obligation to carry pregnancies to term [*Pflicht zur Austragung*]."[69] Nonetheless, the court, in its decision, did endorse another aspect of the law proposed in 1992, one that marked a shift from the prior two options, the trimester-based and the indication-based models, to a new "counseling"-based model for handling abortions. The court signaled that, not least in view of the unmistakable evidence that women's reliance on abortion apparently continued to be quite pervasive, it would permit the development of a law that, while maintaining the *criminality* of abortion, would simultaneously allow an abortion to go *unpunished* if certain conditions were met. It left open what mix of indication and counseling models might be acceptable.[70]

The task now fell to the political parties, and then to the Bundestag as a whole, to propose new versions of the law. Revealingly, the new law proposed by the ruling Christian Democratic / Christian Social Party still included, as though self-evidently necessary, references to the need for an embryopathic indication ("in cases of medical, embryopathic and criminological indication, termination of pregnancy is in accordance with the law"), a sign that the bone of contention for Christian Democrats had all along been the traditional fourth, so-called social indication (the most widely used one and the most contested because it was perceived by antiabortion forces to be inexcusably elastic), which was no longer being mentioned.[71]

Ultimately, however, and whether we read this as a matter either of complete contingency or of multifactor causation, it was in the final hashing-out by a cross-party committee of the various party proposals—a committee that included Christian Democrat Hubert Hüppe, father of a disabled son and a staunch opponent of all abortions, whose minority proposal outlawing abortions on grounds of the "equivalent value of born and unborn life" had already been rejected by the Bundestag—that the embryopathic indication disappeared entirely (to be absorbed, quietly, into the maternal-medical indication).[72] In the small print of commentary on the finally published law, it was explained, tersely, that "for ethical reasons the embryopathic indication has been struck in order to prevent any misunderstanding to the effect that an anticipated disability of a child could be a legitimating basis for a termination."[73]

And so it was that a hardcore antiabortion conservative ended up being the one to give the radical disability rights movement the law it wanted. Despite this manifest victory, moreover, Hüppe's group continued to fret that, potentially, "in the expanded medical indication, terminations on grounds of the disability of an unborn child could be camouflaged"—and it served notice to the legislature and the executive branch of government, via a formal inquiry in 1996, that it continued to be concerned about how the implementation of abortion law was meeting the concern that disabled lives must be valued equally with the nondisabled.[74]

In 2009 these efforts bore fruit. The Bundestag, after prior efforts in 2001 and 2004 had been stalled, formally set yet further restrictions—a three-day waiting period for "reflection" and heightened fines for any doctor discovered to be providing later-trimester abortions because of fetal disability without sufficient proof that bearing the child could

DEUTSCHLAND

NS-Euthanasie: Der Probelauf zum Holocaust

Sie waren die ersten Opfer des systematischen Massenmords der NS-Dikatur: psychisch Kranke und behinderte Erwachsene und Kinder. Die erste EU-Konferenz zur NS-Euthanasie versuchte, Wissenslücken darüber zu füllen.

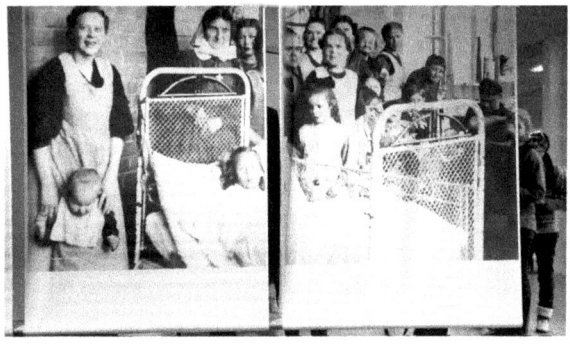

Bild zur Ausstellung "Kinder-Euthanasie"

Eigentlich plante Adolf Hitler bereits 1935 die "Reinigung des deutschen Volkskörpers von psychisch Kranken, Behinderten und Lebensunwerten". Doch noch rechnete er mit zu großen Widerständen in der Bevölkerung. "Die Lösung sollte der Krieg bringen, dann ist das Leben sowieso weniger wert", berichtete Professor Gerrit Hohendorf auf der ersten europäischen Konferenz zu NS-Euthanasie, die drei Tage lang in Berlin stattfand. Viele der 170 Zuhörer aus 20 Ländern erbleichten angesichts der bestialischen Ideologie der Nationalsozialisten, die aus den historischen Zitaten sichtbar wurde.

Hubert Hüppe, Bundesbeauftragter für Behinderte

"Behinderte waren die ersten Opfer des Nazi-Terrors, die organisiert umgebracht wurden - an ihnen wurde geprobt", sagte der Beauftragte der Bundesregierung für die Belange behinderter Menschen, Hubert Hüppe.

"Aktion T4"

Seit dem 18. August 1939 waren Ärzte in Deutschland aufgefordert, Kinder mit Behinderungen den Gesundheitsämtern zu melden. Sie wurden danach in spezielle Anstalten abtransportiert, verhungerten dort oder wurden vergiftet. Der Begriff "Euthanasie" im Sinne von "schönem Sterben" ist also ein falscher Begriff, weil es vielmehr um brutalen Massenmord ging. Der Terminus hat sich aber bis heute in der Geschichtsschreibung etabliert.

Im Oktober 1939 wurde das Tötungsprogramm auf Erwachsene ausgeweitet - und zwar auch in Polen, das im September angegriffen worden war. In der Berliner Tiergartenstraße 4 wurde für die "Aktion T4" eine Verwaltung mit 300 Mitarbeitern aufgebaut. Diese ließen in sechs Kliniken und Anstalten Gasräume einrichten. Hierhin wurden die Menschen mit sogenannten "grauen Bussen" gebracht und dann mit LKW-Abgasen bestialisch umgebracht.

definitively be construed as a threat to the woman's mental health—as Christian Democrats garnered the needed support of Social Democrats and Greens specifically by presenting these amendments to the law as once more an advance for disability rights.[75] From 2009 to 2013 Hüppe served as Chancellor Angela Merkel's federal commissioner for disability issues. From this position he advocated against both stem cell research and preimplantation diagnostics in case of in vitro fertilization.[76] The insight that National Socialist euthanasia was "the trial run for the Holocaust" is part of his self-presentation (see figure 12).[77] In 2012 he joined the Europe-wide One of Us movement—the one that was instrumental in capsizing Portuguese European Union parliamentarian Edite Estrela's "Report on Sexual and Reproductive Health and Rights."

Figure 12 (*left*). Kay-Alexander Scholz, "NS-Euthanasia: The Trial Run for the Holocaust": The Christian Democratic politician Hubert Hüppe, German chancellor Angela Merkel's federal commissioner for disability issues, in 2013 addressed the first European Union–sponsored conference to raise international awareness about the nearly three hundred thousand mass murders of individuals with disabilities and psychiatric illnesses that took place not only within the boundaries of the Third Reich but also in nations occupied by Germany. The conference was held in Berlin and brought together visitors from twenty countries to hear from scholars. The imperative of conducting fresh historical research on the murders in Poland, Hungary, the Czech Republic, France, Austria, and the lands of the former Soviet Union—but also of recovering and translating research that had been conducted in Eastern Bloc lands in the 1960s–1970s but that had been inaccessible to Western experts due to the Cold War—was additionally made clear. Reprinted by permission of Deutsche Welle and picture-alliance.

3

Time Well Wasted

Sexual, Political, and Psychological Subjecthood in the European Union, 2000s–2010s

Disability is not just one difference among many, but a difference that changes everything.

British disability rights feminists
Margrit Shildrick and Janet Price, 2005

We do it slow. What's cool about us is
we take our time
not yours.
Poet Eileen Myles quoted on the cover of the "Queer Methods"
special issue of the journal WSQ, 2016

The twenty-first century has been an auspicious time, so far, for the recognition and even celebration of disability rights. In the geopolitical realm—globally and within the European Union and the United States—and in the realms of media attention, public relations advocacy, and "awareness raising," disability rights have been "in."[1] Billboards on city streets encourage concern and appreciation for difference.[2]

Grassroots advocacy organizations mobilize for every possible cause, with Down syndrome and autism—both understood as forms of cognitive difference—taking the lead, often with either self-confidently defiant or/and sentimental, emotionally engaging slogans.[3] Cinemagoers are treated to public service announcements encouraging friendly contact between abled and disabled individuals.[4] Airports contain signs about the joys of "buddy-dom" between the abled and the differently abled, as well as, of course, especially in the United States, honoring survivors of that other source of disability: warfare.[5] Films, theater productions, and TV shows include story lines with disabled characters (and increasingly now also disabled actors); testimonials and blog commentary fill the internet.[6] News items and documentaries share intimate details about the lives of individuals with physical or cognitive disability (including tales of intensively committed parenting, love stories of adolescents and of adults of all ages, and a huge diversity of "this is what it feels like" self-disclosure).[7] By expanding the visual-aesthetic range of what is offered to the public (perhaps best understood as a very deliberate form of queering, as bodies formerly considered nonnormative become appreciatable, indeed cherishable, and erotically desirable) and by expanding the imaginative-emotional range in offering countless opportunities for identification with different positionalities (the heroically loving, dedicated parent, the self-determining, joyful, differently abled individual, the effective and trusted worker with the laughter-sharing circle of friends, the emulation-worthy savvy consumer, the transgressive or tender lover), the public has been encouraged to a broadened grasp of human possibility and value. This is an enormous and multifaceted collective achievement. It also remains exceedingly fragile.

Meanwhile, there is, almost as though happening on a different planet, a substantial body of twenty-first-century activist and scholarly writings that contend that there was, starting at the latest in the 1980s–1990s—the decades that saw both the massive spread of prenatal diagnostic testing for potential fetal anomalies and eventually the development of preimplantation diagnostics for in vitro fertilization—far from any unlearning of eugenics. Instead, there has been the ascent of a new kind of eugenics, and one that has been spreading ever further in the

2000s–2010s. This new type is all the more insidious, it has been argued, because it has inserted itself into the language of self-determination and freedom of choice, even as, it is additionally averred, both self-determination and freedom of choice must remain suspect concepts in an era that is simultaneously one of the diffusion of neoliberal economic arrangements and ever greater disparities in wealth distribution, dismantling of welfare state supports, and increasing injunctions to individuals to self-entrepreneurialize and become their own risk managers in a world of ever more cutthroat competition and erosion of social solidarities.[8] The promotion of this cluster of interconnected arguments has, perhaps unsurprisingly, been especially prevalent in the German-speaking lands of central Europe, particularly in Germany itself from the early 2000s onward, making its way into the statements of prominent philosophers and cultural critics; of politicians, including in formal parliamentary inquiries; and of health provider organizations.[9] But however salutary the skepticism and caution about the potential perils of new technologies, to focus solely on dramas concerning the beginning of life is to miss the bigger story of the plethora of efforts of individual activists and groups—across European nations—to make persuasive arguments for the rights of disabled individuals already living, to engender empathy, and to model both respect and emotional reciprocity. This chapter can offer just an initial survey of some notable trends, an assemblage of pieces of a far larger and remarkable phenomenon.

Disability, needless to say, is not any one thing but a variously manageable myriad of sometimes enriching, sometimes irrelevant, sometimes devastating vulnerabilities, challenges, and differences from an (itself often fluctuating) norm. There is no question, however, that what the philosopher Hannah Arendt, at one time a stateless refugee, referred to as "the right to have rights" has been extraordinarily elusive for the numerous categories of individuals living with disabilities of one type or another.[10] Of all the many rights enumerated in the Universal Declaration of Human Rights of 1948, and however that declaration's formation is—or is not—understood to be rooted in lessons drawn from the totalitarian, persecutory, and mass murder enterprises of the 1930s–1940s in Europe, identifiable attention to the needs of individuals with disabilities is not there.[11]

It was a nongovernmental organization, the International League of Societies for the Mentally Handicapped (now renamed Inclusion International), that at its meeting in Stockholm in 1967 began to articulate a set of desiderata for securing attention to the specific needs of the cognitively disabled and, at its subsequent meeting in Jerusalem in 1968, formally passed the Declaration on the General and Special Rights of Mentally Retarded Persons.[12] Inspired by this and pushed by advocates, the United Nations took up the theme and in 1971 promulgated the Declaration on the Rights of Mentally Retarded Persons and, four years later, the Declaration on the Rights of Persons with Disabilities— followed by the designation of 1981 as the International Year of Disabled Persons and the decade from 1983 on as the United Nations Decade of Disabled Persons (see figure 13).[13] Yet however impressive an

Figure 13. An official UNESCO calendar with an image of joined candles announcing the United Nations International Year of Disabled Persons 1981 was printed in English, French, and Spanish in conjunction with a major international conference on disability—especially with regard to problems in education for the disabled—held in Torremolinos, Spain. Designed by the Polish artist Jacek Ćwikła and photographed by Marek Koch, the image had won first prize in UNESCO's design competition. The shared flame of the differently shaped candles represented the theme announced for 1981: "Full participation of disabled persons in society." Reprinted by permission of UNESCO and Jacek Ćwikła.

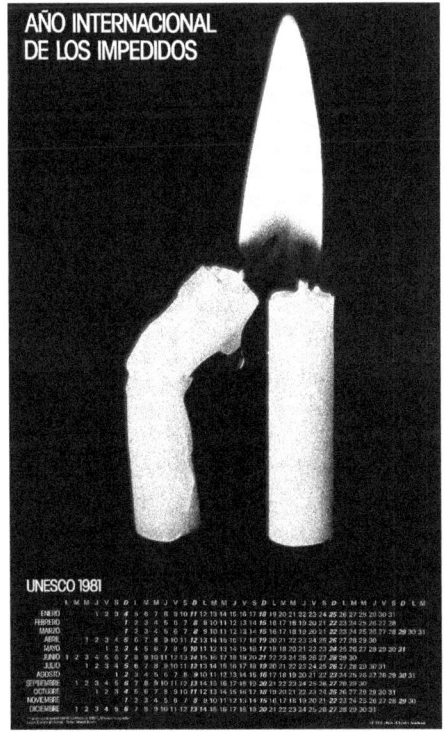

improvement over what had gone before, much of the ensuing activity still saw disabilities, of whatever kind, within the framework of medical and social service needs (often met, if at all, within the context of segregated institutions), rather than of civil rights and of the imperatives to undo exclusion and discrimination. In addition, while declaration followed declaration, and while the efforts to formulate the declarations themselves reveal the constant retractions, modifications, and qualifications necessary to build international consensus, actually putting the demands into practice was stymied at every turn (see figures 14a and b). What was needed was a cultural revolution, a revolution in attitudes, or—as the International League had put it already in 1968—the "creation of a favourable public opinion."[14]

It is this pursuit of a "favourable public opinion" that has had the greatest success since the turn of the millennium. One of the most conspicuous phenomena in the current profusion of activist efforts by and on behalf of individuals with disabilities is that—indicatively, like queer theorists before them—theorists of disability are often pulled back and forth between offering either "minoritizing" or "universalizing" messaging and analyses. Eve Sedgwick, in her pioneering book *Epistemology of the Closet* (1990), had offered the terminological pair "'minoritizing' versus 'universalizing'" as an alternative, when considering the topic of homosexuality, to the more usually invoked but in her view problematic and limited contrast pair "essentialist versus constructivist." Sedgwick's point was not only that there was an irresolvable, but perpetually active, contradiction between "seeing homo/heterosexual definition on the one hand as an issue of active importance primarily for a small, distinct,

Figures 14a (*top right*) and b (*bottom right*). Marked-up notations from the report of working group 4, European Regional Seminar on the International Year of Disabled Persons, held at Siinlinjaärvi, Finland, in May–June 1981. Over and over, qualifying additions or deletions demonstrate how difficult it was to create international consensus on the obligations of states to ensure equality of access to public life and other rights of individuals with disabilities and how quickly optimistic claims were retracted. Reprinted by permission. Location: UNESCO archives, unit ED/SCM B7S1.07-13, box number 1319, UCD 362.4 A 066 "81"/06 (480) "81." Data: ERC Rethinking Disability project (contract no. 648115) database, Leiden University.

...need for parallel systems. Technology has proved that air~~~systems will minimize
~~aft~~, buses, ferries and trains can be made accessible.

RECOMMENDATION 7

27. (a) That existing public transportation systems of each country be accessible. *where possible*

28. (b) That accessibility to terminals, vehicles, and simple information systems in audible and visual form should be ~~introduced~~ at the planning stage for all public transport systems.

29. (c) That people with disabilities should have ~~full access~~ to purchasing, driving or using alternative transport systems ~~of~~ ~~their own choice.~~ *or other citizen.*

30. (d) That the countries should take steps to remove the barrier of extra expense in travelling, i.e. when an accompanying person is needed.

-2-

~~in this respect, for persons with a handicap? Are the~~ *disabled.*
United Nations Declarations followed? Are ~~handicapped~~ persons given equal opportunity to receive education and training?
U Thant declared in 1972, when he was Secretary-General of the United Nations, that if all international and national support for handicapped persons in the Third World was equally distributed, it would amount to one cent a year per handicapped person in developing areas. Even if the amount has increased since then, it clearly shows that countries cannot rely entirely upon external input. Such collaboration can only provide encouragement towards increased national ~~efforts~~. This field ~~of action~~ *Education of disabled persons* must therefore be considered a national responsibility.

2. ~~Special~~ Education facilities *for students with special learning difficulties* are often thought to be expensive in many countries, and this ~~usually means there is a long waiting period~~. A UNESCO study on Economic Aspects of Special Education, however, clearly shows that provisions of education for students with ~~handicaps~~ *disability* is a worthwhile and profitable investment. This judgement is valid for countries in all areas.

3. The title of this topic: Equality in education and training, apparently means equality of access to education and training not equality in results of education...

relatively fixed homosexual minority" (i.e., a minoritizing view) and "seeing it on the other hand as an issue of continuing, determinative importance in the lives of people across the spectrum of sexualities" (a universalizing view in which queerness was understood as part of everyone's life) but also that the contradiction between the minoritizing and universalizing tendencies and the sheer impossibility of adjudicating conclusively between them were hardly minor matters but remained utterly central to "major nodes of thought and knowledge."[15] The dissatisfaction with debating "essentialist versus constructivist" models of homosexuality can be seen as a precursor and parallel to the increasingly felt necessity, within disability theorizing, to find an alternative to the more usually invoked contrast between a medical/charity model based on the idea that there is a deficiency within the disabled individual (the traditional view) and a social constructionist approach (more typical for the 1990s onward) that acts as though the only or at least the main problem is that the social situation is disabling and constraining of the individual—an approach that makes good sense for physical or sensory disabilities (as one can cut curbs for wheelchairs, provide sign language for the deaf, and maybe even adapt one's interactive style to some kinds of neurological difference like high-functioning autism) but that is not always as helpful as one would wish for more severe cognitive disabilities or psychiatric disturbances.[16] Taking a cue from Sedgwick's shift in conceptualization, the comparable point, for disability, might be that what is needed is *both* to attend to the specificities of disabled lives (and not assert glibly that "we are all disabled," because that would be gratuitous nonsense) *and* yet always again also to see the spectrum, the universalizing implications of disability.[17] This tension between the specificities and the universal implications is recurrently evident when European activists work to communicate to a wider public what it would mean to take seriously the "right to have rights" of the disabled across all realms of existence.

Sexual Rights, Sexual Subjecthood

The realm of sex and love has been one of the most crucial sites of transformation in public consciousness around disabilities of almost all kinds. Moreover, significantly, and although they have often gone under-acknowledged, indispensable contributions toward this end were provided by the prior struggles of lesbian, gay, bisexual, and transgender rights movements.[18] Certainly, the sexual revolution itself—the insistence on the morality of pleasure, the refusal or reinvention of normative family forms, and the appreciation for vast individual differences in desire and self-expression—supplied an essential basis. No less important in its wake were the cultural changes at the turn of the millennium that the German sexologist Volkmar Sigusch famously grouped under the rubric of "the neosexual revolution." These include such phenomena as the more frequent dissociation of performed gender from anatomy given at birth; the proliferation and dispersion of what had been infinite private predilections into publicly declared identities; the explosion of media that provide new opportunities for both arousal and connection; the double effect of rising visibility of transsexual bodies as they both confirm and dissolve binary gender ideals; and the diversification and flexibilization of roles of all kinds.[19] All this, taken together, has created a space of openness and receptivity to physical and sexual diversity that is unprecedented. In fact, as we live, globally, through backlashes against and in many places growing ambivalence about the very concept of sexual rights as human rights, it is inside and on behalf of movements for the rights of the disabled to sexual encounters and experiences that some of the strongest arguments for sexual rights as human rights continue to be made.[20]

A case in point was Don Kulick and Jens Rydström's *Loneliness and Its Opposite* (2015), which discussed how sex and disability have been handled in Denmark and Sweden. The authors were resolutely critical of the, as they put it, "patronizing stereotype that disabled adults are like children" and should be kept away from sex, as they saw this as "a furtive way of denying that disabled adults are adults—or even, in an important sense, that they are fully human beings."[21] For them, the need to support and facilitate erotic lives for individuals with disabilities is a moral imperative.

Among their most salient findings was the divergence between national cultures. Kulick and Rydström criticized Swedish approaches, which although frequently preoccupied with politically correct language in reference to disability they found in practice to be expressly sex-negative and repressive, actively denying individuals with disabilities access to sexual and romantic experiences. They saluted and analyzed at great length the opposing approaches prevailing in Denmark. These, as it happened, might well involve un–politically correct language (e.g., "mongol" for Down syndrome and "spastic" for someone with cerebral palsy) but were simultaneously driven by a strong commitment to provide support and assistance to both physically and cognitively disabled individuals so that they could experience sexual contact and pleasures—consistently, and crucially, accompanied by sensitive attention to issues of consent and safety and to distinguishing, rigorously, affection from the risk of abuse. As the authors explained: "The nature of the assistance they receive may well be different, depending on whether the recipient's impairment involves trouble understanding things like why certain behaviors are not allowed in public, or if the impairment involves the absence of or the inability to control movement in legs and arms. But what links [the cases of cognitive and physical disability], despite their differences, is that other people need to engage with and intervene on behalf of the person with the impairment." Drawing on the "capabilities approach" developed by economist Amartya Sen and further by philosopher Martha Nussbaum, Kulick and Rydström contended that "justice is a matter of fostering the circumstances that allow individuals to realize a life with human dignity." And they further invoked Nussbaum as someone who "regards sexuality as a fundamental human entitlement" and who has suggested that "a central feature of a life with dignity is being able to form attachments to other people and having opportunities to develop one's sexuality and seek sexual satisfaction."[22]

Notable as well was the authors' attention to grappling with dilemmas inherent in the Enlightenment legacy for the language of human rights, as especially severer forms of disability pose a challenge to liberal assumptions of bounded individuality, since the assistance needed for these kinds of disability inevitably mean that "personhood is disaggregated" and that "the locus of . . . personhood is dispersed—it resides not in [one] body, but across a network of relations that need to get coordinated." The point of that coordination, however, is precisely *so that* the disabled individual is "able to flourish as an individual." Disability, they further stressed, makes a stronger demand on the nondisabled than queerness makes on the purportedly nonqueer. One of the most splendid things about Kulick and Rydström was their disdain for the often mawkish didactic trend in both media and scholarship that suggests that the disabled can "teach us something." By contrast, Kulick and Rydström noted: "We don't see why they should." They even found the inquiry into "sexual rights" for the disabled, when only posed as a yes/no question, potentially limiting. They wanted instead to redirect the question—and as they did so, the implications were once again universalizing. "'What can we do to help people develop their capability for forming attachments to other people, including attachments that involve sexual pleasure and love?' Phrased like that," they continued, "the question doesn't elicit a yes/no answer; it isn't something a tabloid newspaper can ask its readers to vote on. It isn't even necessarily about people with disabilities. *It is a general question, one that pertains to everybody, and that addresses everybody.* But posed in the context of people with disabilities, it invites a considered engagement with the lives of individuals who need particular kinds of assistance to be able to live a life of dignity."[23] In other words, the authors were calling attention to the interrelatedness and interdependence that are true for all people, but they were also stressing that disability foregrounds that phenomenon with particular acuity, and thereby they once more moved toward the minoritizing pole.

Also in Spain, to take another instance, an organization called Yes, We Fuck has worked actively to promote the message that sexuality is an essential component of a dignified, joyous, full human life. As Spanish sexologist Gemma Deulofeu, who has worked with the organization, has declared: "The disabled have a right to sexual joy." For her, sexuality is "a human right, without exception."[24] Along related lines, in the 2015 film about the organization—directed by Antonio Centeno, an activist in the Independent Living movement, and the documentary filmmaker Raúl de la Morena—a woman (who may be blind) speaks directly to the camera: "Sex is for me one of the most important foundations of the human being, of every person. It's a source of delight, a way of interacting

and it's possible they don't feel much pleasure,

and stay there...pleasure, pleasure and pleasure.

Figures 15a and b. Stills from the film *Yes, We Fuck!* (Spain, 2015). Reprinted by permission.

with people, of personal growth, it's all that for me, and nothing less, and nothing less than that."[25] Tellingly, moreover, it is a man with Down syndrome who articulates for *all* people what is so infinitely remarkable about climaxing. "Listen," he says, "the pleasure is amazing, until you reaches the orgasm, and stay there . . . pleasure, pleasure and pleasure" (see figures 15a and b). And at another moment a different but related example of universalizing relevance is expressed by two women who appear in the film. They belong to a group that markets adapted sex toys and also does provocative street theater so as to explode what counts as sexual (again a form of queering). In the film, they are addressing *all* people when they say: "You have to have imagination to fuck. And most people have been told that 'this' [i.e., the traditional unimaginative way] is fucking, and it's possible they don't feel much pleasure. But they don't know that there may be other ways to do it."[26]

Over and over, then, the tension between universalizing and minoritizing messages is especially apparent in regard to sexuality. Thus, to give one more example from 2014–15, in Germany a man named Christian Bayerlein, who happens to have a disintegrative muscle disease, lost his job as the disability coordinator for the city of Koblenz because he had given an interview to the left-leaning newspaper *taz* in Berlin about his romantic life, which included polyamory and S/M practices, as well as long-term loves. Both the interview itself and the dismay felt by progressives that the Christian Democrats in Koblenz were apparently glad to have him working on access ramps in the city but not pleased with him divulging details of his personal life had universalizing implications. But unmistakably evident too was the minoritizing message that Bayerlein was professionally vulnerable due to his disability because he was stepping out of the bounds of the appropriately pitiable and into proud defiance.[27]

Political Rights, Political Subjecthood

In short, there is yet more to the challenge that disability poses, and this is even clearer when we our shift focus from sexual subjecthood to political subjecthood. The key complication is that, as a number of disability theorists have tried to put into words, "disability is not just one difference among many [i.e., it is *not* like race or class or gender], but a difference that changes everything." As British disability feminists Margrit Shildrick and Janet Price put it in 2005, "Disability is not just another add-on concern."[28] So too, and also in 2005, the French psychoanalyst (and government-appointed disability rights spokeswoman) Julia Kristeva has argued that disability creates "*this exclusion that is not like others.*"[29]

What makes it so? Kristeva's argument was that there was something viscerally threatening about disability—whether physical or cognitive— and this was what made it not comparable to race, religion, class, or sexual orientation. Disability, as she put it, "confronts the able-bodied person with the limits of life, with the fear of deficiency and physical or psychical death: disability therefore awakens a catastrophic anxiety that in turn leads to defensive reactions of rejection, indifference, or arrogance"—if not also "the will to eradicate by euthanasia." Repeatedly, she circled back to the point that "the disabled person opens a *narcissistic identity wound* in the person who is not disabled; he inflicts a threat of *physical or psychical death*, fear of collapse, and, beyond that, the anxiety of seeing the very *borders of the human species* explode. And so *the disabled person is inevitably exposed to a discrimination that cannot be shared*." Or again, in another variant: "Those not afflicted with these incapacities are faced with the anxiety of castration; the horror of narcissistic injury; and beyond that, the intolerableness of psychical or physical death." Disability, moreover, she reminded her reader, was not necessarily about improvability or progress—to say nothing of cure—but rather all too often stasis and stagnation or even worsening.[30] In short, in the confrontation with disability, there is often a different relationship to time, to futurity, to the relationship between effort and outcome.

Kristeva, meanwhile, criticized France in particular for having been late to the cause of justice for the disabled—an outrage, she noted, for a nation that prided itself on being the country of the "rights of man." Riffing on the French Enlightenment thinker Denis Diderot, who in 1749 famously wrote the *Letter on the Blind for the Use of Those Who See*, she too wrote the (Diderot-echoing) "Open Letter to the President of the Republic on Citizens in a Situation of Handicap for the Use of Those Who Are and of Those Who Are Not." In it she argued for "a true cultural revolution, based on real interaction between abled and disabled in such realms as school, business, culture" and a shift from seeing the disabled as "'objects under treatment'" to understanding them as *emerging subjects*."[31]

Ultimately, in trying to reach the heart of the reader, she drew as well on her experience as a psychoanalyst and made a universalizing point that she hoped could be the basis for overcoming, as she put it, the "abyss" that still separates the world of the abled from the world of the disabled. It is on the psychoanalytic couch, she noted, that the vulnerability that is in *everyone* becomes apparent, and she proposed adding a fourth term to the trio "Liberty, Equality, Fraternity," and that was "Vulnerability." So rather than arguing, as some disability activists do, that one should care about the disabled because "'it could happen to anyone,'" she suggested instead that we should care because "*it* [disability] is already in me/us: in our dreams, in our anxieties, our romantic and existential crises, in this lack of being that invades us when our resistances crumble and our 'interior castle' cracks." Because, in other words, in our deepest being, we are all already vulnerable. She continued: "Contrary to the propaganda with which globalized technology assaults us, the global age that has followed the modern age is not one of the high-performance, pleasure-filled Man, bisexual master of his desires, their debacles, or both. The vulnerability that is revealed today on the couch is precisely that which is determined to deny the manic invasion of hyper-productivity, all-pervasive spectacle, and suicidal religious warfare." So her hope was that to recognize that vulnerability in all of us, "in me," as she put it, that is, in herself, "will help me discover the incomparable subject in the limited body," that is, to discover the "emerging subject" inside the disabled person. On this basis she dreamed that the visceral fear and sense of narcissistic injury triggered by contact with disability could be "transformed into attention, patience, and solidarity," and that in view of the recognition of the universality of vulnerability, the disabled could even help to transform the abled.[32]

Others worked on the issue of political subjecthood in more pragmatic and concrete ways. A crucial document was the UN Convention on the Rights of Persons with Disabilities. Decided December 2006 and effective May 2008, it was the first international human rights convention of the twenty-first century and—strikingly—the convention with the shortest history of negotiations and the highest number of national signers in the first two years of its existence. It was also the first international human rights convention ratified by the European Union as such.[33] The development and ratification of the convention were the outgrowth of many decades of engaged activism, prepared for by the various prior declarations but representing both a fresh conception and a new strength of commitment at the highest levels of many governments. The convention represented a third step beyond the medical individual-deficiency model and the subsequently developed social disablement-by-the-environment model.[34] Starting from the social paradigm, which sees individuals with disabilities as subjects with rights, not as objects of charity, the convention went yet further. Now the charge was to state governments actively to transform the environment in accordance with the principles of nondiscrimination, equality, and inclusion. While in 1980–81, when the United Nations International Year of Disabled Persons was in preparation, it was evident already in the planning stages that goals would not be met and thus the language of the idealistic demands was constantly being watered down and modified, *this* time the aim was to see to it that the rights that were declared could truly be realized.[35] The convention contains sections on the special concerns of disabled women (recognizing also the chilling fact of their greater vulnerability to abuse, as well as the specific needs of disabled women who are mothers) and disabled children (here the convention calls for equal rights with other children, the preeminence of the child's well-being, and the guarantee of the child's freedom of opinion in all matters concerning it). There are mechanisms in the convention for checking up on countries' compliance; there is an oversight committee that is composed of individuals with disabilities (in other words, the disability movement slogan "Nothing about us without us" is meant to be put into practice); and there is a strong global development component (relevant not least because two-thirds of the 650 million people with disabilities worldwide live in developing countries).[36]

One of the most important figures in the development of the language of the convention and thereafter in liaising with the German state over its implementation was Theresia Degener, an internationally renowned lawyer and disability activist from the German radical "cripple movement" and one of the "thalidomide children" born in the early 1960s.[37] Degener has also been crucial in explaining what the big breakthroughs in the convention were. So, for instance, in explaining how the convention started from a social model ("the fact that someone uses a wheelchair or thinks or feels more slowly or differently from others does not cause the exclusion from society, discrimination and the withholding of rights do") but did yet more than that, Degener noted that the convention was expressly an "empowerment convention."[38] Violations of rights were now recognized as such rather than being "trivialized as inescapable individual fate," and—even more importantly—it was now the positive duty of the state to do away with discrimination.[39]

Degener, too, but from a different vantage from Kristeva, thought about the Enlightenment ideals of liberty and equality and how they needed to be supplemented and/or rethought for the case of disabilities. Thus, for instance, Degener called attention to Article 12 of the convention, which innovated with its idea of "supported decision-making" in situations involving law or property—a way, in short, to reconcile disabled individuals' needs for help and care with respect for their right to be individually self-determining. While Kristeva revisited Diderot, Degener built on the bioethicist Sigrid Graumann, who retooled Immanuel Kant as she formulated the concept of "assisted freedom."[40] (We can note that the message here, while articulated for the disabled minority, was once again universalizing. For of course *all* our freedoms are assisted freedoms, since interdependence is how reality actually works. But it is the disabled who make this phenomenon palpable.) At the same time, it remains crucial to register that "assisted freedom" cannot occur without cost. Disability assistance—whether provided in shifts or round-the-clock—is all too often low-paid, strenuous, stressful work, and an articulation and addressing of assistance workers' own rights to dignity, decent pay, and good working conditions is just now coming onto the horizon.[41]

The concept of "assisted freedom" has recently been especially discussed with respect to the issue of voting rights, those quintessential democratic political rights, and how they could be realized also for—as the official European Union phrasing put it—"persons with intellectual disabilities" and "persons with mental health problems."[42] Here again were conceptually universalizing implications, as those who objected that cognitively disabled or psychiatrically ill people could not know what is best for them, or for anyone else, or even understand what was at stake at all were now rebuffed by those who noted just how many

voters who were assumed to be *non*disabled regularly chose candidates
that others wished they would not.

When asked whether he did not worry that "people with intellectual
disability could be manipulated in their voting decisions," Valentin
Aichele of the German Institute for Human Rights (Deutsches Institut
für Menschenrechte) retorted, "That would be a problem that also affects
other people." First, not just because, with mail-in ballots, for instance,
one did not know who actually filled them out but because human beings
in general were all too often manipulable: "After all, no other citizen
who is entitled to vote is required to lay bare his reasons for this vote.
Some are led by sympathies, others by family traditions or solely by the
content of the political program."[43] Second, and in any event, the advocates for voting rights emphasized that every individual deserved representation of his or her specific interests as he or she, together with those
he or she trusts, perceived them.[44]

In many countries those who are cognitively disabled or psychiatrically ill were long denied the right to vote in their nation's elections. This
situation has now come under pressure due to the UN Convention—
not least in Germany, where approximately ten thousand individuals
have been excluded from voting because they have comprehensive
guardianship (also in financial and medical decision-making).[45] Some
countries, like Austria, have never had this barrier; others, for instance,
the UK in 2006 and the Netherlands in 2008, have deliberately elimi-
nated it quite recently, seeing it as a human rights violation.[46] As of 2010
the European Union Agency for Fundamental Rights had made voting
rights for the disabled and psychiatrically ill a formal agenda item and—
in conjunction with Inclusion Europe, a suborganization of the group
now called Inclusion International that had first raised the issue of the
rights of individuals with cognitive disabilities in 1967—between 2010
and 2017 produced a series of publications on legal and political rights
for the disabled written in accessible "easy read" language.[47]

Life-Sharing Models, Psychological Subjecthood

Yet the capacity to vote once or twice a year in various elections, or to
have one's "right to have rights" affirmed by supranational and national

governmental authorities, hardly means being experienced and treated, every day, as someone with a full psychological personhood. Here, a renewed return to an earlier history is necessary to rediscover sources of stimulus and orienting practical models in the past—especially the ones that continue to animate life-sharing projects and the newest ways of theorizing and living with disability in our present juncture. As material and grounded in quotidian interactions as the intentional communities that were launched during the war or in the postwar decades were, they too invariably involved strategies of persuasion to engage a wider public, and they too demanded innovative reflection on the relationships between ability and disability of various kinds. Interestingly—importantly—these experiments came in both religious (whether quasi- or deeply so) and explicitly secular variants. And—no less importantly—a number of them had their roots in opposition to Nazism and all that it stood for.[48] A focus on these experiments provides yet another conceptual genealogy for beginning to answer the question of how Europeans have worked to unlearn eugenics.

One example is the Camphill movement, founded by Karl König, a Jewish Austrian émigré to Scotland. Camphill was based in the semi-Christian anthroposophical principles of Rudolf Steiner—better known to most people as the impetus for the holistic, experience-based education model followed by the Waldorf schools worldwide, but also someone who had (in 1924) developed an alternative pedagogy for the disabled called "curative education," including rhythmic movement exercises, arts, massage, and therapeutic biodynamic farming, as well as a kind of cosmic spirituality that saw in every person, no matter how disabled, a soul that was whole. König's way of putting it was to see in the children "the spark of the living spirit . . . present in each one of them in spite of their deficiencies." A Viennese pediatrician who had received training at anthroposophical medical centers in Switzerland and Germany and treated many children with cognitive disabilities, König turned out to have a special gift for working with these children, and when he returned to Austria, he continued to provide services for them. In the 1930s he gathered around him in Vienna a group of mostly Jewish, mostly socialist-leaning, young, idealistic intellectuals who were intrigued by the teachings of Steiner and who wanted to live in accordance with

anthroposophical principles. Peculiar as some of the principles were—
and one usefully could compare them, for instance, with those of
Steiner's contemporary, the breakaway psychoanalyst Carl Jung, who
had his own loopy cosmic spiritual sides—the fact is that in practice
they facilitated quite tender care for individuals with disabilities. As
König once put it, "We do not want to read Anthroposophy; we want to
live it. We decided to aim at starting a home for handicapped children."[49]

Moreover, concretely, the Camphill movement was born out of the
double fact that König's group had to flee Vienna when the Nazis took
over Austria in 1938 and that the group, reunited in Scotland, committed
itself to living in intentional community with cognitively disabled individuals as an express counter to the murderous onslaught against the
disabled that was already being adumbrated and then was to be set in
motion in practice in the Third Reich just one year later. (One historian
has described Camphill as "an anti-fascist alternative to concentration
camps," "a kind of inverse reflection of Nazism."[50]) Another major influence on König was the Welsh utopian socialist Robert Owen, founder
of the cooperative movement. Having fled individually, when they re-
grouped in an isolated rustic house in the countryside near Aberdeen
donated to them by a British anthroposophist contact, six weeks later
they were joined by the first disabled children; many more soon followed,
sent by desperate parents, as, also in the United Kingdom, institutional
conditions for the disabled, under the impact of the eugenics movement,
were horrendous.[51] The first child was the son of German refugees fleeing
to the United States who, due to US immigration restrictions denying
entry to the disabled, could not take their son with them. Immediately,
the group was confronted with the severity of cognitive disability, as
this boy seemed "like a little beast[.] He cannot talk, wipes the food
from the table with his tongue, and pokes his hands and his nose in
every pocket to look for empty cigarette boxes, which are his only toys."
Within a year, in 1940, as crowding in the first house became a problem, the director of the Macmillan publishing company, himself the
father of a disabled son who needed care, bought the house, Camphill
Estate, that became the first formal home for the fledgling movement.
By 1949, four years after the war's end, further houses had been built,

and the original community had 183 children, with a waiting list of 123 more; by 1957 there were 272 pupils and 124 coworkers.[52] Today, there are over one hundred Camphill communities in twenty nations.[53]

A second initiative in shared living was the deeply Catholic but now increasingly ecumenical movement called L'Arche (as in the French word for Noah's ark). L'Arche was started in Trosly-Breuil, France, in 1964 by the Catholic Canadian philosopher (and former student for the priesthood) Jean Vanier, who was spurred to this work through his friendship with a priest, Thomas Philippe, who had called his attention to the outrageously bad conditions for the disabled in institutions in the postwar period. But Vanier was additionally influenced by his experiences, as he was the son of the Canadian ambassador to France in 1945, seeing and working to help the emaciated and terrified survivors of Nazi concentration camps. Vanier soon took the position that the disabled are the teachers of the nondisabled.[54] Again there were universalizing implications, as L'Arche has understood itself to this day as modeling a way of living that can be inspirational for all. Certainly there are now also Protestant, Jewish, and Muslim religious voices insisting on rethinking theology from the perspectives of the disabled.[55] Yet L'Arche is increasingly making its argument in more general humanistic terms. There is a broadly felt sense—articulated frequently by individuals involved in or moved by L'Arche—that as the world reels ever more into ugly crises, different forms of interhuman interaction need to be practiced.[56] Among L'Arche's express goals are "the development of long-term, mutual, interdependent relationships" and "the maintenance of a stable, life-giving home environment" but also "to make known the gifts of people with intellectual disabilities." Vanier—by now a much-revered figure on the international stage—insists he has been helped more than he has helped.[57] This is, indeed, a crucial thematic, repeated often by people involved in the movement (see figure 16).[58] The idea is that the nondisabled *need* the disabled. There are now 149 L'Arche communities around the world in thirty-eight nations.[59]

A third model, no longer extant but the focus of quite a revival of interest and fascination in the last couple of years, is that of the Frenchman Fernand Deligny. Deligny's completely secular experiment in living

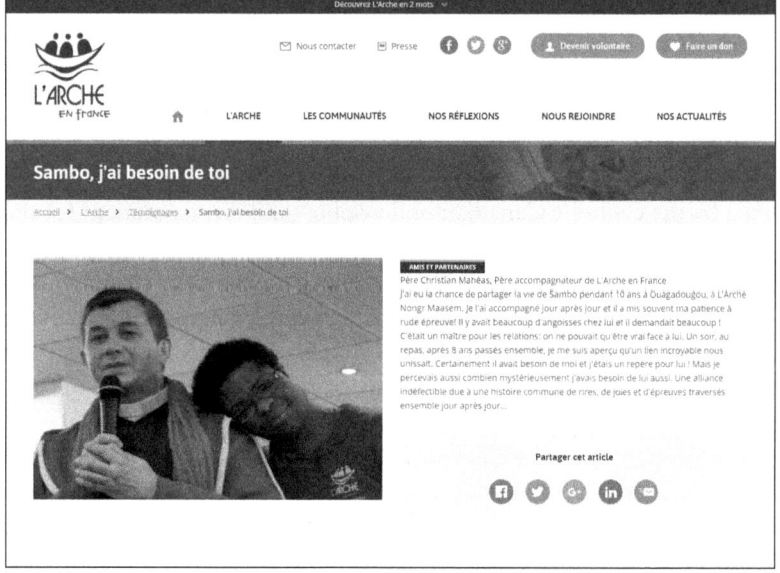

Figure 16. "Sambo, I need you." Father Christian Mahéas writes about the ten years he spent at a L'Arche community in Ouagadougou, Burkina Faso, and the "incredible bond" and "unwavering alliance" that united him particularly with Sambo, one of the residents. Reprinted by permission of L'Arche en France.

in intentional community with autistic and nonverbal children in the Cévennes mountains in the 1950s–1970s was rooted not least in his work in the French resistance to German Nazi occupation in the 1940s. Another tributary source lay in the related project of alternative psychiatric care, also based in shared living, provided for the mentally ill during the German occupation by the Catalan psychiatrist François Tosquelles at the clinic of Saint-Alban, subsequently furthered by French psychiatrist Jean Oury at the clinic of La Borde, and joined in the postwar period by the young leftist psychoanalyst Félix Guattari.[60]

Living with severely challenged children in a barren landscape, Deligny began a practice of tracing their movements, creating images he called "wander lines" (*lignes d'erre*) and eventually filming them, together with documentarian Chris Marker (with whom he had worked since 1946); he was also close to François Truffaut.[61] One of the most compelling of these films is called *Le moindre geste* (The slightest gesture).[62] The name itself captures Deligny's strong attunement to the personhood of the children he had rescued from a life within oppressive institutions to bring into a kind of, in Deligny's words, "primordial communism."[63] While Jacques Lacan, the French psychoanalyst whose seminars Deligny had attended, had been preeminently focused on language and on meaning, Deligny found this supremely unhelpful as he mused on living with children deemed to be "outside of speech." Instead of attempting to train the children, Deligny and his coworkers found themselves "in search of a mode of being that allowed them [the children] to exist even if that meant changing our own mode." As a scholar recently noted, Deligny was "anticipating, by several decades, some of the central tenets of the neurodiversity and autistic self-advocacy movements."[64] Rather than—as psychoanalytic approaches, unfortunately, so often had—fixating on sourcing a disability (often blaming parents), offering only the insight that the nondisabled harbored "death wishes" on the disabled, and/or attempting perpetually to *interpret* (usually by imposing meaning on) the strange behaviors or unintelligible utterances of disabled children, Deligny emphasized *being-with* them, seeing himself as living within "a network" whose "project is to bring autistic children into close contact" and indeed whose "true project" might best be seen as "the network itself, which is a mode of being."[65] And even as he played with the image of a spider weaving a web, he deemphasized his own agency, writing instead of how "the web's project is to be woven"— and at another point that "if the spider indeed sought out the nook, we may also say that the nook was waiting."[66]

A Desire to Live Otherwise

For secular theory, activism, and disabled-abled life sharing in the present, however, no historical touchstone has been more important than the recovery and extension of the work of Félix Guattari. Somewhat like Deligny (albeit also differently—since at the La Borde clinic founded by Jean Oury, Guattari lived and worked with schizophrenic adults rather than mute and cognitively disabled children, and since he was far more overtly engaged with leftist political mobilization), Guattari too focused on *being-with*, indeed on what he called *becoming*, and he certainly believed that human beings were mutually transformative. Indicatively, moreover, Guattari has been the focus of a huge renaissance these past few years within, notably, both queer studies and disability studies.

Guattari's ideas, one might say, were queer *avant la lettre*. He is best known as the coauthor, with philosopher Gilles Deleuze, of the 1972 counterculture classic, *Anti-Oedipus: Capitalism and Schizophrenia*, and its sequel, *A Thousand Plateaus*.[67] Deleuze and Guattari developed a blend of the ideas of the British analyst Melanie Klein (who saw inner chaos in everyone, and a rich stew of fantasies) with those of Wilhelm Reich (who was preoccupied with why human beings were so often split within themselves, drawn both to authoritarianism and to antiauthoritarian rebelliousness but in general far more frequently to the political Right than the Left). They challenged liberal ideas of the boundedness of individuals and instead saw humans as composed of crazy swirling insides, interconnecting with parts of everyone and everything else in a constant flux of flow and stoppages. They believed human beings could go fascistic-paranoid or liberatory-generous at any given moment. They famously challenged all ideas of identity boundaries, insisting on a blur between white and black, hetero and homo, man and woman—indeed, they recommended (and this was a radical form of universalism) that ideally all human beings might engage in "becoming-black," "becoming-gay," or "becoming-woman." And they saw desire not as chasing some lack but as a positive and productive force.[68]

For physical disability queer theorists Margrit Shildrick and Janet Price—mentioned already before—the beauty of Deleuze and Guattari was that their ideas "queer the nature of individuality" (see figure 17). They noted that such challenges to autonomous bounded individuality were relevant for everyone, giving the example of how an adaptive car augments a disabled person both literally, as a "phenomenological extension of [a] body," and conceptually, as in a "rhizomatic proliferation of connections." Workers in France, Thailand, or the Philippines

Figure 17. "Deleuzian Connections and Queer Corporealities," *Rhizomes* website, Fall 2005–Spring 2006, accompanying the article by British disability rights feminists Margrit Shildrick and Janet Price. Reprinted by permission of Margrit Shildrick and *Rhizomes: Cultural Studies in Emerging Knowledge.*

coproduce what makes a disabled body in the UK expand its functions, even as, in turn, these workers' lives are changed by participation in this network of assemblages. Invoking sex toys or running shoes, Shildrick and Price observed: "The point is not that disabled people are unique in relying on a profound interconnectivity, but that for the normative majority such a need, inevitable as it is, may be covered over—particularly in the domain of western hegemony." But they also said that recognizing the ubiquity of interconnectivity could make the disabled seem less pitiable and marginal: "Once the focus switches from separation to connection, . . . a corporeal mode that has figured only as a devalued deficiency must be reassessed. The disabled woman who relies on an assistant or carer to help her prepare for a sexual encounter—be it in terms of dressing appropriately, negotiating toilet facilities, or requiring direct physical support in a comfortable sexual position—is not different in kind from other women. . . . Similarly a reliance on prosthetic devices—the linkages between human, animal and machine—would figure not as limitations but as transformative possibilities of becoming." And again they argued: "It is difficult to draw a meaningful distinction between, for example, a disabled person's use of a voice synthesiser and the growing use of text messaging, between reliance on a hearing dog and riding a horse, between the amputee who uses a wheelchair cart and the executive in a speedboat, or between the wearer of corrective orthopaedic boots and an athlete's high-tech sprint shoes. All extend the bounds of possibility by making connections—by both organic and technological means." To underscore the point, they quoted disability theorists David Mitchell and Sharon Snyder: "The prostheticized body is the rule, not the exception." And, as they noted, citing science scholar Donna Haraway, what was going on in all these instances was "queering what counts as nature."[69]

Disability scholars working with individuals with cognitive disabilities—though verbal—have of late drawn on Deleuze and Guattari as they sought to do what is called "inclusive research" (in other words, research *with* rather than *on* or *for* disabled people). Thus, for instance, Griet Roets and Dan Goodley, in Belgium and the UK, saw the individuals who were organized in People First self-advocacy groups as "politicized citizen subjects" who should be listened to, since they have a lot to say about "the mundane incivilities and ontological violations that are part of disabled people's everyday lives." As one of their informants remarked as he rejected the pathologizing label of intellectual disability: "We do not want that label anymore. Certainly, you people have the control. You are the experts in your field, but I am an expert too, in my own field. I lived in the institution, and I lived in a group home."[70]

Finally, however, and in conclusion, one of the most remarkable Guattarian experiments has been lived and reflected on in Malta, the smallest country in the European Union. Here the education specialists Duncan Mercieca and Daniela Mercieca have worked with children with PMLD (profound and multiple learning disabilities)—many of them, like Fernand Deligny's mountain children, entirely nonverbal and often incapable of any other standard form of communication. The severity and multiplicity (often physical in addition to cognitive) of these children's difficulties have led the Merciecas to theorize beyond not only the medical and social models but also the emancipatory/empowerment model exemplified by the UN Convention and its related "assisted freedom" and "supported decision-making" concepts.

The Merciecas have found the social model, even especially in its explicit "emancipatory" mode, as not transformative and not thoughtfully conceived enough. They have been interested in the most deeply disabled persons, the children with PMLD, and how these children can transform the researchers, allowing the researchers to be changed by not just being-with these children but by becoming-with them, turning these nonverbal subjects of research into agents who can affect the lives of the nondisabled. The social model, they have noted, assumes that the task of the researchers is "disclosure of the reality out-there and hence emancipation of the disabled." But, following from Deleuze and Guattari, the Merciecas have called for research itself to be "rethought as becoming through engagement with intensities rather than as interpretation of a pre-existent reality out there." For them, "emancipating and empowering" are something done "to others" and hence something that "retains echoes of the charity model that the social model deprecates." The idea instead is "to affect and be affected." Moreover, "the researcher needs to desire the engagement." Such research, they hope, "gives space to what is not measurable, to what is left out, to what is considered as nonsense"; it "allows for contingency and a lack of certainty." Yet "this kind of research is political in that the researcher places herself on the margins of the research community, and at the limits of herself, and in so doing has the space to express another potential community . . . the invention of a possibility of human being to come."[71]

In his recent follow-up book, *Living* Other*wise* (2013), Duncan Mercieca provided concrete examples of the intensities he and others (nondisabled youths and adults alike) have experienced in their interactions with children with PMLD, how these children have been agents who have transformed him, and how he *desires* to be with them, deeply longing for their company (see figure 18). And, indeed, how he thinks—like Guattari did about becoming-black or becoming-gay or becoming-woman—about the profundity of the possibility of "becoming-PMLD." These are children with an IQ range of below 20, whose skillset is that of an eighteen-month-old, who are totally physically reliant on others for their basic care, and who have an "unrefined sense of cause and effect." (In other words, the people Peter Singer did not and could not see as "persons.") For Mercieca, these *are* persons who, he writes, make him think—indeed, "think again." Invoking the standard notion that "success is identified with saving time," therefore, "thinking has a fatal flaw of making us waste time," Mercieca notes that "to think *again*, then, will be to waste time twice over," but he thinks this thinking-again that the children provoke in him is thus not just "wasting time" but "time well wasted." Acknowledging "the intellectual paralysis that often surrounds educational staff working with students with PMLD" and "the stagnation which may be present in situations involving [their] care," he sees in the children that they "make us experience a lack in our own thinking; a deficiency in our thinking and living. But at the same time they offer us a possibility to fill this lack, to live beyond what we are living now." To desire to be with these students, he suggests in his conclusion, "does not mean a desire of lack, a desire of what you do not have, but a desire of connection . . . an enjoyment of flows of matter and energy . . . a desire of experimentation . . . [a] desire for the other, a desire to live *otherwise*, a desire for the student with PMLD."[72] In sum, then, if we traverse the paradigm shifts over the course of the postwar decades in engagements with disability, the move has been, on the whole, one from *charity* to *justice*. But the most provocative recent efforts, several of them returning to Deligny or Guattari, involve more than justice; they also are based on *desiring* and *becoming*.

A final image to consider involves the Merciecas' idea of "time well wasted." It is exemplified by textual artwork displayed on the cover of a recent special issue, published in 2016, titled "Queer Methods" of the

STUDIES IN INCLUSIVE EDUCATION

Living *Other*wise

Students with Profound and Multiple Learning Disabilities as Agents in Educational Contexts

Duncan P. Mercieca

*Sense*Publishers

Figure 18. Cover of Duncan P. Mercieca, *Living* Other*wise* (2013). Reprinted by permission.

American gender and sexuality studies journal *WSQ*. The text is taken from a poem by the poet Eileen Myles that is included in the issue. Obviously playing on slow and sensuous lovemaking—but interpretable as about slower forms of thinking or moving as well—the artwork reads: "We do it slow." And then it continues, even more cleverly: "What's cool about us is / we take our time / not yours."[73] So while this graces the "Queer Methods" issue, the message would fit as well for opening oneself up to the intensities, flows, and forces of "becoming-disabled." One way of unlearning eugenics, then, might be to see this too as not solely a minoritizing message but also a potentially universalizing one . . . a message that could speak to everyone.

Acknowledgments

I want to thank especially Michael Staub, Moshe Sluhovsky, Todd Shepard, Ilana Löwy, Karina Korecky, and Caroline Arni, who each, at key moments, has helped me think in new ways about a number of the most troubling and difficult issues raised here. I have benefited greatly as well from wonderfully engaged audiences at Johns Hopkins, Columbia, the University of Chicago, and the University of Massachusetts, Amherst. Above all, it was an extraordinary honor to deliver these chapters first as the George L. Mosse Lectures at the Hebrew University in Jerusalem in December 2016. I had met Mosse at the "Lessons and Legacies" conference of the Holocaust Educational Foundation at Dartmouth in 1994 and have a vividly positive recollection of the encounter. While at first glance the subjects of disability and abortion may seem far afield from his interests, actually the ways I have sought to address the complex problems these subjects raise—about bodies and rights, "health" and aesthetics, LGBT and other sexual politics, Nazism and its aftermath, and cultural conflict more generally—are very much inspired by his work and intended as a tribute to his memory. The lectures' format and style have been retained here; the note apparatus is purposely capacious. The hope is that readers will find there a trove of evidence and questions to pursue further.

Notes

Chapter 1. Abortion and Disability

1. On these points, see also Dagmar Herzog, "The Death of God in West Germany: Between Secularization, Postfascism, and the Rise of Liberation Theology," in *Die Gegenwart Gottes in der modernen Gesellschaft: Transzendenz und religiöse Vergemeinschaftung in Deutschland*, ed. Michael Geyer and Lucian Hölscher (Göttingen: Wallstein, 2006), 425–60. A prior, longer version of one portion of the material in this chapter first appeared under the title "Christianity, Disability, Abortion: Western Europe, 1960s–1970s," in *Archiv für Sozialgeschichte* 51 (2011): 375–400. The material is reprinted here with kind permission of the Friedrich-Ebert-Stiftung e.V.

2. Martine Sevegrand, *Les enfants du bon Dieu: Les catholiques français et la procréation au XX⁰ siècle* (Paris: Albin Michel, 1995); Angelo Somers and Frans van Poppel, "Catholic Priests and the Fertility Transition among Dutch Catholics, 1935–1970," *Annales de Démographie Historique* 2 (2003): 57–88; Crystel Hug, *The Politics of Sexual Morality in Ireland* (New York: Palgrave Macmillan, 1999).

3. Alana Harris, ed., *The Schism of '68: Catholicism, Contraception and "Humanae Vitae" in Europe, 1945–1975* (New York: Palgrave Macmillan, 2018).

4. Harry Oosterhuis, "Christian Social Policy and Homosexuality in the Netherlands, 1900–1970," *Journal of Homosexuality* 32, no. 1 (1996): 95–112; Gert Hekma, *Homoseksualiteit in Nederland van 1730 tot de moderne tijd* (Amsterdam: J. M. Meulenhoff, 2004), 100–113; Frédéric Martel, *Le rose et le noir: Les homosexuels en France depuis 1968* (Paris: Seuil, 1996); Albrecht D. von Dieckhoff, *Der Griffin-Report* (Hamburg: Decker, 1956).

5. Susan Gal, "Gender in the Post-socialist Transition: The Abortion Debate in Hungary," *East European Politics and Societies* 8, no. 2 (1994): 256–86; Agnieszka Graff, "We Are (Not All) Homophobes: A Report from Poland," *Feminist Studies* 32, no. 2 (2006): 434–49.

6.   See the remarks of the editor of the conservative paper *Il Foglio*: Giuliano Ferrara, "Dimenticare la 194 e combattere l'aborto," *Il Foglio*, May 20, 2008, http://www.ilfoglio.it/articoli/2008/05/22/news/dimenticare-la-194-e-com battere-l-aborto-73114/; see also "Church in Italy Insists on Abortion Moratorium," *Catholic News Agency*, January 30, 2008, http://www.catholicnews agency.com/news/church_in_italy_insists_on_abortion_moratorium/; "Abortion Big Issue in Italy Elections," *CBSNews.com*, February 12, 2006, http://www.cbsnews.com/news/abortion-big-issue-in-italy-elections/; Hilary White, "Excitement, High Hopes for Brussels' 1st Ever March for Life," *LifeSiteNews*, March 24, 2010, https://www.lifesitenews.com/news/excitement-high-hopes-for-brussels-st-ever-march-for-life.

7.   Italy will be discussed in detail below. For Spain and Portugal, see James Markham, "In Spain, Feminism Clashes with Tradition," *New York Times*, April 11, 1980; José Linhard, "Family Planning in Spain," *International Family Planning Perspectives* 9, no. 1 (1983): 9–15, here 13; Marvine Howe, "Abortions in Portugal—a Complex Controversy," *New York Times*, March 13, 1976; Jill Jolliffe, "Abortion Passes, Coalition Teeters in Portugal," *Globe and Mail*, January 28, 1984.

8.   For a revelatory transnational overview of the high prevalence of illegal abortion in the first post–World War II decade (along with information about declining rates of posttermination maternal mortality due to abortionists' increasing use of penicillin), see Karl-Heinz Mehlan, "Abortstatistik und Geburtenhäufigkeit in der Deutschen Demokratischen Republik," *Das deutsche Gesundheitswesen* 19 (1955): 1648–59, esp. 1657–58. The Western European countries mentioned are Denmark, Sweden, France, England, Austria, Finland, and West Germany.

9.   The Legislative Decree of May 31, 1946, no. 561, was directed against "newspapers or publications or other printed matter that divulge the means meant to impede procreation, or that illustrate the use of them, or that give indication on the mode of procuring them, or that contain inserts or correspondence relative to the said means" and announced that such publications "can be sequestered." See Giancarlo Matteotti, "Proposta di Legge," *Camera dei Deputati*, July 23, 1958.

10.   Quoted in Hans Harmsen, "Mittel zur Geburtenregelung in der Gesetzgebung des Staates," in *Sexualität und Verbrechen*, ed. Fritz Bauer et al. (Frankfurt am Main: Fischer, 1963), 183. Along related lines, Simone Veil in France in 1974 was acutely conscious that many parliamentarians were worried

about the French birthrate, and she rushed to assure them that decriminalizing abortion would not damage the birthrate further.

11. Theodor Bovet, *Von Mann zu Mann: Eine Einführung ins Reifealter für junge Männer* (Tübingen: Katzmann, 1955), 47.

12. Marcus Collins, *Modern Love: An Intimate History of Men and Women in Twentieth-Century Britain* (London: Atlantic, 2003), 173.

13. Luigi De Marchi and Maria Luisa Zardini, "Bringing Contraception to Italy: Pathfinder in Italy," in *Courageous Pioneers: Celebrating 50 Years as Pathfinder International and 80 Years of Pioneering Work in Family Planning and Reproductive Health*, ed. Linda J. Suttenfield et al. (Watertown, MA: Pathfinder International, 2007), 39.

14. Maud Dugrand and Mina Kaci, "Trente ans après, Simone Veil se souvient," *L'Humanité*, November 26, 2004, http://www.humanite.fr/node /316453.

15. Harmsen, "Mittel zur Geburtenregelung," 175.

16. Letter of March 20, 1950, quoted in Judith G. Coffin, "Sex, Love, and Letters: Writing Simone de Beauvoir, 1949–1963," *American Historical Review* 115 (October 2010): 1061–88.

17. On Switzerland, see Stefan H. Pfürtner, "Moralwissenschaftliche Erwägungen zur Strafrechtsreform des Schwangerschaftsabbruches," in *Zur Frage des Schwangerschaftsabbruches: Theologische und kirchliche Stellungnahmen*, ed. Hermann Ringeling and Hans Ruh (Basel: Reinhardt, 1974), 42–67, here 45. For the UK, see "Abortions in Britain Total 40,000 a Year," *The Guardian*, July 15, 1966, 5 (of the forty thousand in this estimate, three-quarters were said to be illegal); and—for the higher estimate—see Church Assembly Board for Social Responsibility, *Abortion: An Ethical Discussion* (Westminster: Church Information Office, 1965), 7. On West Germany, see Angela Delille and Andrea Grohn, *Blick zurück aufs Glück: Frauenleben und Familienpolitik in den 50er Jahren* (Berlin: Elefanten, 1985), 123; "Anti-Baby Pillen nur für Ehefrauen," *Der Spiegel*, February 2, 1964, 87; and Carl Nedelmann, "Abtreibung: Geburtenregelung und Strafrechtsreform," *Konkret*, July 1965, 6. For the three million or more estimate for Italy, see Rosemary Ruether, "Italy's 'Third Way' on Abortion Faces a Test," *Christianity and Crisis*, May 11, 1981; and Arthur Marwick, *The Sixties: Cultural Revolution in Britain, France, Italy, and the United States, c. 1958–c. 1974* (Oxford: Oxford University Press, 1998), 713. For lower numbers (eight hundred thousand to one million a year), see Massimo Livi-Bacci, "Demografia dell'aborto in Italia," *Sapere*, no. 784 (1975): 41–46; Patrick

Hanafin, *Conceiving Life: Reproductive Politics and the Law in Contemporary Italy* (New York: Routledge, 2007), 29.

18. Veil quoted in "Der Widerschein meiner persönlichen Überzeugungen," *Die Weltwoche*, December 4, 1974, 3.

19. "Ich habe nur Umgang mit Mörderinnen," *Der Spiegel*, May 31, 1971, 134, 136.

20. Reprinted in German translation in Michaela Wunderle, *Politik der Subjektivität: Texte der italienischen Frauenbewegung* (Frankfurt am Main: Suhrkamp, 1977), 103.

21. Public self-accusation campaigns occurred in France, Germany, Italy, and Spain. In the Spanish case, thirteen hundred women self-accused in response to a trial of thirteen women in Bilbao—the self-accusers pointed out that the only difference between them and the (poorer) defendants was that they had had the financial means (40,000 pesetas) to travel to England to get abortions. "'Yo he abortado voluntariamente,' declaran mil trescientes mujeres," *El País*, October 20, 1979.

22. *Pardon* cartoon accompanying the essay: "Abtreibung: Massenmord oder Privatsache?," *Der Spiegel*, May 21, 1973, 38–50.

23. Paul Ferris, "What Abortion Reform Means," *The Guardian*, July 17, 1966, 9.

24. "Abortion Law Reformers Attacked," *The Guardian*, February 14, 1967, 8.

25. Methodist conference quoted in Malcolm Potts et al., *Abortion* (New York: Cambridge University Press, 1977), 295.

26. Hermann Ringeling, "Fragen um den Schwangerschaftsabbruch," in Ringeling and Ruh, *Zur Frage*, 41.

27. Gyula Barczay, "Für die Fristenlösung," in Ringeling and Ruh, *Zur Frage*, 94.

28. Church Assembly Board for Social Responsibility, *Abortion*, 61.

29. Cf. Ruether, "Italy's 'Third Way' on Abortion Faces a Test"; Beverly Wildung Harrison, "Our Right to Choose," in *Women's Consciousness, Women's Conscience: A Reader in Feminist Ethics*, ed. Barbara Hilkert Andolsen et al. (San Francisco: Harper & Row, 1987). But similar arguments were made also in Switzerland. See especially Barczay, "Für die Fristenlösung," 101.

30. "Pour une réforme de la législation française relative à l'avortement," *Études*, no. 338 (1973): 55–84.

31. Eberhard Jüngel, Ernst Käsemann, Jürgen Moltmann, and Dietrich

Rössler, "Abtreibung oder Annahme des Kindes: Thesen zur Diskussion um den Paragraph 218," *Evangelische Kommentare*, no. 8 (1971): 452–54.

32. Barczay, "Für die Fristenlösung," 97.

33. Pfürtner, "Moralwissenschaftliche Erwägungen," 49. Pfürtner was of German heritage, a member of the Dominican order from 1945 to 1974, working in the late 1960s and early 1970s in Switzerland; he was later named a Righteous among the Nations at Yad Vashem for having hidden and saved three Jewish women from the concentration camp Stutthof, hiding them in his parents' home in Danzig, during the Third Reich. The Nazis had at one point arrested him, supposedly for mocking Hermann Göring, then released him for time served after six months in isolation—his military officer had intervened on his behalf—after which he went back to the Eastern Front. He became a theologian after the war. Already two years before this essay on abortion, he had written an important book about Christianity and sexuality, calling for liberalization: *Kirche und Sexualität* (Reinbek: Rowohlt, 1972). The book was received by the Catholic hierarchy as subversive of its authority, and so he was asked to leave his professorship, which he did; he also left the Dominican order in 1974.

34. Church Assembly Board for Social Responsibility, *Abortion*, 7.

35. Pfürtner, "Moralwissenschaftliche Erwägungen," 52.

36. Church Assembly Board for Social Responsibility, *Abortion*, 17, 24–25, 28–29.

37. Barczay, "Für die Fristenlösung," 102.

38. Bishop of Essen Franz Hengsbach and Catholic physician Siegfried Ernst both quoted in "Abtreibung: Massenmord oder Privatsache?," 39.

39. *Neue Bildpost* quoted in "Ich habe nur Umgang."

40. "Compte rendu integral—2e séance du mardi 26 novembre 1974," *Journal Officiel*, November 27, 1974, http://archives.assemblee-nationale.fr/5 /cri/1974-1975-ordinaire1/071.pdf. Interestingly, in post-Franco Spain a few years later, references to German Nazism and the murder of the disabled were included by a conservative author opposing any liberalization of Spanish abortion law. Gonzalo Nadal, "Cartas al director: Nacer: El primer derecho del niño," *ABC*, March 28, 1979, 79.

41. For a terrific analysis of the simultaneous trends—with similar intertwining of disability and abortion debates—in the United States in the 1960s, see Leslie J. Reagan, *Dangerous Pregnancies: Mothers, Disabilities, and Abortion in Modern America* (Berkeley: University of California Press, 2010). For a revelatory assessment of the broad popular support for liberalization of abortion

access in the United States and of the multiple factors eventually causing shifts in party and political alignments around the issue, see Linda Greenhouse and Reva B. Siegel, "Before (and after) Roe v. Wade: New Questions about Backlash," *Yale Law Journal* 120, no. 8 (June 2011): 2028–87.

42. Ben Quinn, "Thalidomide Victims Get Apology from Makers after Half a Century," *The Guardian*, August 31, 2012, http://www.theguardian.com /society/2012/sep/01/thalidomide-victims-get-apology-from-grunenthal.

43. The Thalidomide UK Agency has reported that "there are 458 people currently in the UK who were affected by the drug, but that for every thalidomide baby that lived there were 10 that died." Recently, it has been reported that in addition to the survivors, "about 6,000 were miscarried, 2,000 were stillborn and a further 2,000 died in infancy." See James Meikle, "Thalidomide 'Caused up to 10,000 Miscarriages and Infant Deaths in UK,'" *The Guardian*, March 6, 2016, https://www.theguardian.com/society/2016/mar /06/thalidomide-caused-up-to-10000-miscarriages-infant-deaths-uk. See also Harold Evans, "Thalidomide: How Men Who Blighted Lives of Thousands Evaded Justice," *The Guardian*, November 14, 2014, https://www.theguardian .com/society/2014/nov/14/-sp-thalidomide-pill-how-evaded-justice; and the website of the Thalidomide UK Agency, http://www.thalidomideuk.com/. Evans states: "The original catastrophe maimed 20,000 babies and killed 80,000: war apart, it remains the greatest manmade global disaster."

44. A countervailing, or at least complicating, argument—that, not least because thalidomide had tended to be ingested by "respectable" middle-class women, the scandal surrounding the ensuing malformations helped legitimate abortion more generally and direct attention away from its prior associations with irresponsibility and "promiscuity," while the ensuing chatter about "monstrosities" was profoundly insensitive and could even well be described as *immoral*—can be found in brilliant essays analyzing the impact of the scandal across the British Commonwealth, including Australia, New Zealand, Canada, and South Africa. Notably, too, in these contexts there could be found religiously inspired arguments not just against but also in favor of abortion rights; there were in addition recurrent invocations of Nazi eugenics and mass murders of the disabled. See Clare Parker, "From Immorality to Public Health: Thalidomide and the Debate for Legal Abortion in Australia," *Social History of Medicine* 25, no. 4 (2012): 863–80; Susanne M. Klausen, "'There Is a Row about Foetal Abnormality Underway': The Debate about Inclusion of a Eugenics Clause in the Contraception, Sterilisation, and Abortion Act, 1977–1978," *New Zealand Journal of History* 51, no. 2 (2017): 80–103; Christine

Chisholm, "The Curious Case of Thalidomide and the Absent Eugenic Clause in Canada's Amended Abortion Law of 1969," *Canadian Bulletin of Medical History* 33, no. 2 (2016): 493–516; and Susanne M. Klausen and Julie Parle, "'Are We Going to Stand By and Let These Children Come Into the World?': The Impact of the 'Thalidomide Disaster' in South Africa, 1960–1977," *Journal of Southern African Studies* 41, no. 4 (2015): 735–52.

45.   Following the explosion and knowledge of the toxic cloud "infecting" the area, 604 women in Seveso asked for advice and help at clinics; of these, 414 were pregnant and 192 were within the first ninety days of pregnancy (within which they could ask for a "therapeutic abortion" in accordance with a 1975 law). Twenty-six underwent abortions.

46.   "Ich habe nur Umgang," 141.

47.   "Pour une réforme."

48.   Bernhard Häring, *Heilender Dienst: Ethische Probleme der modernen Medizin* (Mainz: Grünewald, 1972), 99.

49.   Ringeling, "Fragen um den Schwangerschaftsabbruch," 26–27.

50.   Rickie Solinger, "The Population Bomb and the Sexual Revolution: Toward Choice," in *American Sexual Histories*, ed. Elizabeth Reis (Oxford: Blackwell, 2001), 342–75.

51.   Cf. Dagmar Herzog, *Sexuality in Europe: A Twentieth-Century History* (Cambridge: Cambridge University Press, 2011), esp. 18–27.

52.   Auguste Forel, *Die sexuelle Frage* (Munich: Reinhardt, 1909), 504.

53.   Hildegart Rodríguez quoted in Alison Sinclair, "The World League for Sexual Reform in Spain: Founding, Infighting, and the Role of Hildegart Rodríguez," *Journal of the History of Sexuality* 12, no. 1 (2003): 98–109, here 104.

54.   Topic Collection 12, box 14, file A (SxMOA1/2/12/14/A), no. 028, May 1949, and no. 029, May 1949, Mass Observation Archive, University of Sussex, Special Collections at The Keep. See in this context also the extensive documentation of the persistence of eugenic thinking and its imbrication with the postwar welfare state in Clare Hanson, *Eugenics, Literature, and Culture in Post-war Britain* (New York: Routledge, 2013).

55.   Nowhere was this more disturbingly apparent than in postwar West German culture, where the strong support the postwar medical community gave to those physicians who had participated in the mass murder of the disabled was palpable—and where contempt for the disabled continued to be expressed, also among self-defined Christians. Ernst Klee, *Behinderten-Report* (Frankfurt am Main: Fischer, 1974); Klee, "'Turning the Tap On Was No

Big Deal': The Gassing Doctors during the Nazi Period and Afterwards,"
Dachau Review 2 (1990); Götz Aly, *Die Belasteten: "Euthanasie" 1939–1945—
Eine Gesellschaftsgeschichte* (Frankfurt am Main: Fischer, 2013); Sascha Topp,
*Geschichte als Argument in der Nachkriegsmedizin: Formen der Vergegenwärti-
gung der nationalsozialistischen Euthanasie zwischen Politisierung und Historiogra-
phie* (Göttingen: Vandenhoeck & Ruprecht, 2013).

 56. See especially the work of the European Union Agency for Fundamental Rights (FRA), "People with Disabilities," http://fra.europa.eu/en/theme
/people disabilities. For older but very useful assessments, see Gerard Quinn
and Theresia Degener, *Human Rights and Disability: The Current Use and
Future Potential of United Nations Human Rights Instruments in the Context of
Disability* (New York, 2002); and Anna Lawson and Caroline Gooding, eds.,
Disability Rights in Europe: From Theory to Practice (Oxford: Hart, 2005).

 57. See, for example, Ann Furedi, "'Disability Cleansing'—or a Reason-
able Choice?," *Spiked*, August 29, 2001, http://www.spiked-online.com/newsite
/article/11261#.V_gfcCQo-Uk; Carine Vassy, "How Prenatal Diagnosis Became Acceptable in France," *Trends in Biotechnology* 23, no. 5 (May 2005); the
debates in Germany in Gisela Notz, "Guter Tag für 'Lebensschützer,'" *SoZ—
Sozialistische Zeitung* 6 (2009): 6; and "Gentests an Embryonen: 'Es gibt keinen
Dammbruch,'" *Spiegel Online*, July 13, 2010, http://www.spiegel.de/wissenschaft
/medizin/gentests-an-embryonen-es-gibt-keinen-dammbruch-a-705997.html;
the Italian cystic fibrosis case is discussed in "European Court Ruling Creates
'Right to Eugenics,'" *Life Site News*, August 31, 2012, https://www.lifesitenews
.com/news/european-court-ruling-creates-right-to-eugenics; and the arrival of
the trends in the United States in Stefanija Giric, "Strange Bedfellows: Antiabortion and Disability Rights Advocacy," *Journal of Law and the Biosciences* 3,
no. 3 (2016): 736–42. The most recent focus for alarm among abortion opponents is the news from Iceland. See, for instance, Lorie Johnson, "The 'Holocaust' against Down Syndrome Unborn Babies," *CBNNews*, March 21, 2017,
http://www1.cbn.com/cbnnews/health/2017/march/world-down-syndrome
-day-celebration-marred-by-hococaust-in-iceland-elsewhere; Lauren Bell, "In
Iceland 100% of Babies Diagnosed with Down Syndrome Are Aborted. Think
about That," *Life Site News*, March 14, 2017, https://www.lifesitenews.com
/opinion/babies-with-down-syndrome-deserve-love-not-eradication; and Sheila
Gunn Reid, "Iceland's President Supports World Down Syndrome Day—Then
Does THIS to Solve 'Problem,'" *TheRebel.media*, March 21, 2017, https://
www.therebel.media/iceland_s_president_supports_world_down_syndrome_
day_but_does_this_to_solve_problem_of_down_s_babies. The latter report
declared that Iceland is "committing a genocide," with "public health programs

to exterminate people before they're born," and it went on to say that "it's the kind of eugenics that Hitler used," except that "luckily" his evil had not been "wrapped . . . up in a women's march and a knit pussy hat."

58.   See the press release of Pro Familia, "Rückschritt im Abtreibungs-recht," May 15, 2009:

> True assistance for women, who after the twelfth week of pregnancy decide for a termination, which they will only receive with a medical indication, will not be provided by this new change in the law. To say that it will is nothing but hypocritical pretense. We ask instead, what difficulties will result from the change in the law for affected women in the future? Here it is important to dif-ferentiate between women who are pregnant in their thirteenth week and women who are, after the major ultrasound and later, pregnant after the twenty-second week. It will not be a relief for women to be subjected to a fixed period of days in order "quietly" to be able to think about their decision— what an ignorant, contemptuous image of women lies behind such a concept! They will have three days of fear to worry about whether the doctor will grant them a medical indication. Also the doctor gains three days to reflect on whether he wants to subject himself to the risk of providing a—possibly contestable— medical assessment, additionally threatened with a fine of 5.000 Euro if found guilty. He will tend only then to provide the medical indication if the patient is in danger of actually losing her life. This division of the medical indication is a definitive setback for women's health politics.

See http://www.profamilia.de/?id=2461 (accessed July 31, 2011).

59.   For the summary text concluding the Parliamentary Inquiry into Abortion on the Grounds of Disability (July 2013), see http://dontscreenusout .org/wp-content/uploads/2016/02/Abortion-and-Disability-Report-17-7-13 .pdf. For criticism of the inquiry, see Ann Furedi, "Abortion: 'We Can Trust Women to Make Decisions That Are Right,'" *Telegraph*, January 31, 2013, http://www.telegraph.co.uk/women/womens-life/9840067/Inquiry-into -abortion-law-on-disabled-babies-We-can-trust-women-to-make-decisions -that-are-right-says-BPAS-chief-Ann-Furedi.html. For an emphatic defense, see "End Discrimination against Unborn Disabled Children Says US Inquiry," *SavingDownSyndrome*, July 17, 2013, http://www.savingdownsyndrome.org/end -discrimination-against-unborn-disabled-children-says-uk-inquiry/.

60.   Critics were outraged that the "malformation" (*malformación*), "dis-ease" (*enfermedad*), or "anomaly" (*anomalía*) of a fetus were being assimilated into the United Nations' protected category of "disability" (*discapacidad*) against which discrimination was impermissible. María R. Sahuquillo and Vera Guttiérez Calvo, "Las verdades a medias para limitar el supuesto de

malformación: El PP justifica su reforma en la presión de la ONU y de colectivos de discapacitados," *El País*, December 20, 2013, https://elpais.com/sociedad /2013/12/19/actualidad/1387483571_951701.html. The proposed law used the phrase "*anomalías incompatibles con la vida*" as the only legitimate cause for termination. In demanding protection for fetuses, Alberto Ruiz Gallardón spoke of "*algún tipo de minusvalía o de malformación.*" The law can be found at http://www.unav.edu/documents/58292/004aaf94-5e5a-4a14-84a2 -4ae8574b387a; the minister of justice interview is with Francisco Velasco, "La malformación del feto no será ya un supuesto para abortar," *La Razón*, July 22, 2012, http://www.larazon.es/historico/5803-alberto-ruiz-gallardon-la-malfor macion-del-feto-no-sera-ya-un-supuesto-para-abortar-LLLA_RAZON _475528. See also Teresa Lamas, "Spain's New Abortion Law," *Women News Network*, February 5, 2014, https://womennewsnetwork.net/2014/02/05 /spains-new-abortion-law/; and Luis Losado Pescador, "Spain's New Abortion Restrictions Are Actually a Betrayal of the Government's Pro-life Base," *Life Site News*, September 10, 2015, https://www.lifesitenews.com/opinion/spains -new-abortion-restrictions-are-actually-a-betrayal-of-the-governments.

61. Małgorzata Fuszara, "Legal Regulation of Abortion in Poland," *Signs* 17 (1991): 117–28; Joanna Goven, "Gender Politics in Hungary: Autonomy and Antifeminism," in *Gender Politics and Post-Communism: Reflections from Eastern Europe and the Former Soviet Union*, ed. Nanette Funk and Magda Mueller (London: Routledge, 1993), 224–40; Agnieszka Graff, "The Land of Real Men and Real Women: Gender and E.U. Accession in Three Polish Weeklies," in *Global Empowerment of Women: Responses to Globalization and Politicized Religions*, ed. Carolyn M. Elliott (New York: Routledge, 2008), 191–212.

62. One of the most recent entries into the newfound concern for disability—joined with demographic worries—can be seen in Belarus in 2016: Lizaveta Kasmach, "Pro-life vs Pro-choice in Belarus," *Belarus Digest*, October 10, 2016, http://belarusdigest.com/story/pro-life-vs-pro-choice-belarus-27507.

63. The text of "Hungary's Constitution of 2011" is available in English at https://www.constituteproject.org/constitution/Hungary_2011.pdf.

64. Kata Janecskó, "Hiába védett a magzat, nem szigorodik az abortusz," *Index*, March 11, 2011, http://index.hu/belfold/2011/03/11/az_abortuszrol_tor veny_valtozasa_varhato.

65. Puppinck "Memorandum on the New Hungarian Constitution of 25 April 2011," May 19, 2011, http://www.eui.eu/Documents/General/Debating theHungarianConstitution/MemorandumontheNewHungarianConstitution .pdf. For a countervailing (but ultimately unsuccessful) feminist legal perspective

on the Hungarian amendment, see the commentary written by the Hungarian Women's Lobby and the Center for Reproductive Rights: http://www.reproduc tiverights.org/sites/crr.civicactions.net/files/documents/CRR_HWL_Consti tution_comments_Hungary_March_2011FINAL.PDF. For another critical analysis of the Constitution as a whole, see http://lapa.princeton.edu/hosted docs/amicus-to-vc-english-final.pdf.

66. On the voting down of "the bill to ban abortion for foetal malforma-tion" in Poland in September 2013, see Colin Francome, *Unsafe Abortion and Women's Health: Change and Liberalization* (London: Routledge, 2016), 38.

67. See Rick Lyman and Joanna Berendt, "Poland Steps Back from Stricter Abortion Law," *New York Times*, October 6, 2016; and "Anti-abortion Po-land Offers Payments for Disabled Newborns," *Fox News World*, November 4, 2016, http://www.foxnews.com/world/2016/11/04/anti-abortion-poland -offers-payments-for-disabled-newborns.html. One reason for the focus on disability is that it remains one of the very few grounds on which abortion is permitted at all in Poland. Of the only 1,040 abortions performed legally in Poland in 2015 (others are performed illegally, or women travel to Slovakia, Germany, or other nations), it is estimated that "many are linked to Down syndrome." Andrew Roth, "'Her Story Is My Story': How a Harsh Abortion Ban Has Reignited Feminism in Poland," *Washington Post*, November 18, 2016. The British *Guardian* ran an article explaining that the law would have made women who terminated their pregnancies for any reason punishable with up to five years in prison. It also quoted statements from protesters against the proposed law who deemed it "barbaric" and "really cruel," noted that resist-ance to the law was about "basic dignity of a woman," and worried that "they will add to current legislation that foetuses with Down's syndrome are not severely damaged so pregnancies with those children won't be allowed to be terminated." One woman quoted observed that "people do not understand how this legislation would affect the lives of Polish families. They do not under-stand the lives of families with disabled children. Raising a disabled child in Poland is very hard. Benefits are just 35 euros a month." See Carmen Fishwick, "'It's about Basic Dignity': Six Women on Protesting Poland's Anti-abortion Proposal," *The Guardian*, October 6, 2016, https://www.theguardian.com/world /2016/oct/06/its-about-basic-dignity-six-women-on-protesting-polands-anti -abortion-proposal. The British *Daily Mail* provided additional numbers: "Government figures say 1,040 abortions were performed in Poland last year, while experts say some 150,000 abortions a year are done illegally and secretly." "Anti-abortion Poland Offers Payments for Disabled Newborns," *Daily Mail*,

November 4, 2016, http://www.dailymail.co.uk/wires/ap/article-3905382
/Anti-abortion-Poland-offers-payments-disabled-newborns.html#ixzz4
XvXkeL8T. For an effort to defend, from a Catholic theological perspective,
the retention of a woman's right to decide to terminate a pregnancy on grounds
of disability, to explain that in some cases such a termination could be the
more moral choice, and to encourage the Catholic bishops of Poland to understand the deleterious effect on physicians in Poland already of current law—as
many of the physicians are even afraid to provide prenatal testing for fetal ab-
normalities for fear of prosecution—see "Academic Urges Polish Bishops to
Support 'Early, Safe and Legal' Abortion for Disabled Babies: Prof Tina Beattie
Has Signed a Letter to Polish Bishops Opposing Plans to Ban Abortion in the
Country," *Catholic Herald*, April 25, 2016, http://www.catholicherald.co.uk
/news/2016/04/25/academic-urges-polish-bishops-to-support-early-safe-and
-legal-abortion-for-disabled-babies/.

68. Politics in the European Union works these days via a complex interaction of supranational, nation-state, NGO, and citizens' initiative actors. For
the Estrela Report itself, see http://www.europarl.europa.eu/sides/getDoc
.do?pubRef=-//EP//TEXT+REPORT+A7-2013-0426+0+DOC+XML+V0//
EN.

69. On the rising phenomenon of right-wing NGOs, see Neil Datta,
"Keeping It All in the Family: Europe's Antichoice Movement," *Conscience* 34,
no. 2 (2013): 22–27; Christopher McCrudden, "Transnational Culture Wars,"
International Journal of Constitutional Law, April 1, 2015, 434–62; and Elena
Zacharenko, *Perspectives on Anti-choice Lobbying in Europe: Study for Policy
Makers on Opposition to Sexual and Reproductive Rights and Health in Europe*
(November 2016), http://www.heidihautala.fi/wp-content/uploads/2017/01
/SRHR-Europe-Study-_-Elena-Zacharenco.pdf.

70. See this critique of the Estrela Report and its supporters by European
Dignity Watch: "Victory: European Parliament Stands for Human Dignity.
ESTRELA Report Not Adopted," October 22, 2013, http://www.europeandig
nitywatch.org/estrela-report-not-adopted/. European Dignity Watch has been
referred to by the European Humanist Federation as "one of the most prominent and active anti-human rights NGOs operating at the EU level." See European Humanist Federation, "European Dignity Watch," https://humanistfed
eration.eu/radical-religious-lobbies/european-dignity-watch/. On HazteOír,
see J. Lester Feder, "The Rise of Europe's Religious Right," *BuzzFeedNews*,
July 28, 2014, https://www.buzzfeed.com/lesterfeder/the-rise-of-europes-reli
gious-right?utm_term=.ru879°W1Zq#.wnN5dlw6DK.

71. For HazteOír, see, for example, "Mienten," *HazteOír.org*, December 21, 2016, http://citizengo.org/hazteoir/ed/39730-mienten. Here as elsewhere there is a big concern that "the gay group" wants to enforce "sexual indoctrination of children" via the insistence on "nondiscrimination" and thereby "silence our free voices." One of the most effective strategies mobilized against the Estrela Report was the idea that it would—through its emphasis on sex education—encourage "premature sexualization of children." Rumors were avidly circulated that the Estrela Report would encourage kindergartners to masturbate. See Pierre-Arnaud Perrouty, "Dangerous Liaisons: The Extreme Religious Lobby at the European Union," *Conscience*, April 25, 2016, 24–29; and Stefan Niggemeier, "Der Deutschlandfunk über 'Gleichschaltung' und 'Homosexualität als Leitkultur,'" January 14, 2014, http://www.stefan-niggemeier .de/blog/17010/der-deutschlandfunk-die-gleichschaltung-und-homosexuali taet-als-leitkultur/. For brilliant assessments of the larger context in which LGBT and reproductive rights—as well as access to reproductive technologies— are assaulted by new configurations of both secular and Catholic conservatives, see Camille Robcis, *The Law of Kinship: Anthropology, Psychoanalysis, and the Family in France* (Ithaca, NY: Cornell University Press, 2013); and Robcis, "Liberté, Égalité, Hétérosexualité: Race and Reproduction in the French Gay Marriage Debates," *Constellations* 22, no. 3 (2015): 447–61.

72. Kuby quoted in European Humanist Federation, "European Dignity Watch." Although the NGO is registered in Belgium, the president and founder of EDW, Jorge Soley Climent, is Spanish, and Kuby is German.

73. On One of Us, see http://www.oneofus.eu/.

74. On ECIs and One of Us's singular success in comparison with other citizens' initiatives, see Elsa Hedling and Anna Meeuwisse, "The European Citizens' Initiative Stage: A Snapshot of the Cast and Their Acts," in *EU Civil Society: Patterns of Cooperation, Competition and Conflict*, ed. Håkan Johansson and Sara Kalm (New York: Palgrave, 2015), 210–28. Initially, One of Us was rejected by the European Commission in its bid to stop funding for embryonic stem cell research. See "European Citizens' Initiative: European Commission Replies to 'One of Us,'" http://europa.eu/rapid/press-release_IP-14-608_ en.htm. However, for its central role in the defeat of the Estrela Report, see "Resolution on Sexual and Reproductive Health and Rights (1): The LGBTI (2) and Pro-abortion Lobby Thwarted by European Citizens," *Gènéthique*, October 1, 2013, http://www.genethique.org/en/resolution-sexual-and-repro ductive-health-and-rights1-lgbti2-and-pro-abortion-lobby-thwarted-61881.

75. On Puppinck, see Order of St. Andrew the Apostle, Archons of the

Ecumenical Patriarchate, "Gregor Puppinck," http://www.archons.org/con
ference/bio-puppinck.asp. On Sekulow, see American Center for Law and Justice, "About Jay Sekulow," http://aclj.org/jay-sekulow. On the ECLJ, see
http://eclj.org/. On the ACLJ's original founding by Pat Robertson, see Feder,
"The Rise." On US evangelical involvement in "religious freedom" and other
cases brought to the European Court of Human Rights, as well as on the
broadly influential role of Puppinck and the European Centre for Law and
Justice, see Pasquale Annicchino, "Winning the Battle by Losing the War: The
Lautsi Case and the Holy Alliance between American Conservative Evangeli-
cals, the Russian Orthodox Church and the Vatican to Reshape European
Identity," *Religion and Human Rights* 6 (2011): 213–19. Annicchino has summarily described the ECLJ as "a Conservative Christian pro-life law firm, associated to the American Center for Law and Justice" (216). On the ACLJ's involvement in Europe—and more generally on current battles to redefine the
meanings of "religious freedom" in Europe, see Ronan McCrea, "Singing from
the Same Hymn Sheet? What the Differences between the Strasbourg and
Luxembourg Courts Tell Us about Religious Freedom, Non-discrimination,
and the Secular State," *Oxford Journal of Law and Religion* 5 (2016): 183–210.

76. On "neonatal infanticide," see ECLJ, "Stop Neonatal Infanticide!,"
https://eclj.org/abortion/un/stop-neonatal-infanticide; on "transgender marriage," see ECLJ, "Transgender Marriage up Next?," https://eclj.org/transgender
-marriage-up-next; and for an example of how surrogacy is understood as a violation of human rights, see the report "Surrogate Motherhood: A Violation of
Human Rights," presented at the Council of Europe on April 26, 2012, http://
www.ieb-eib.org/en/pdf/surrogacy-motherhood-icjl.pdf.

77. Puppinck listed the following "values" enshrined in the Constitution

- The Rejection of the National Socialist and Communist Dictatorship
- The Nation based on ethnic origin
- The Reaffirmation of the Underlying Christian Values of the Hungarian
  State and Society
- The Cooperation between Church and State
- The protection of the right to life and human dignity from the moment of
  conception
- Protection of the family and the institution of heterosexual marriage
- The condemnation of practices aimed at eugenics

At another moment he observed: "While some radical pro-LGBT lobbies may
not appreciate the new Hungarian constitution, it is the sovereign and legitimate
choice of the Hungarian people. This choice, which includes protecting life

from conception, is especially justified by the serious demographic problem that Hungary suffers regarding an exceedingly low fertility rate—around 1.3 children per woman" (Puppinck, "Memorandum," 4, 17).

78. "European court ruling creates 'right to eugenics.'"

79. Grégor Puppinck, "The More Precautious Approach of the ECHR," https://eclj.org/eugenics/echr/human-rights-and-eugenics-the-more-precau tious-approach-of-the-echr. Another antiabortion website says that thirty Down and disability organizations supported the protest. See "Epilogue of the AK Affair versus Latvia: A Down Syndrome Child Does Not Constitute 'Prejudice,'" http://www.genethique.org/en/epilogue-ak-affair-versus-latvia-down -syndrome-child-does-not-constitute-prejudice-61676.html#.WKj19X8°-Uk. This website states: "Under the banner of 'Stop Eugenics Now to protect disabled people / Stopper l'eugénisme maintenant pour protéger les personnes handicapées!,' a petition, supported primarily in France by the Fondation Jérôme Lejeune, received over 10,000 signatures." On the phenomenon of "wrongful birth" cases, see Albert Ruda, "'I Didn't Ask to Be Born': Wrongful Life from a Comparative Perspective," Journal of European Tort Law 1, no. 2 (2010): 204–41; Ewa Bagińska, "Wrongful Birth and Non-pecuniary Loss: Theories of Compensation," Journal of European Tort Law 1, no. 2 (2010): 171–203.

80. See "1st ONE OF US Forum and 1st ONE OF US Award," http://righttolife.org.uk/news/success-one-us-pan-european-forum/; as well as "Success of 'One of Us' European Forum," http://righttolife.org.uk/news/success -one-of-us-pan-european-forum/. A Bulgarian website—proudly offering "non-religious reasons to advocate for life"—also reported on the Pan-European Forum and summarized the challenges that One of Us sees itself as responding to: "Eugenics, research on embryo, gestational surrogacy, euthanasia, transhumanism: there are many challenges faced by this new force for life and human dignity in Europe." "Първи форум на европейската федерация 'ЕДИН ОТ НАС,'" Избор за живот, https://www.pro-life.bg/en/2016/03/1st -one-of-us-forum-and-1st-one-of-us-award/. The story of Baby Gammy is complicated, and versions conflict. One of Us praised Chanbua for refusing to abort Gammy, but actually it appears that it was only after the twins were born that the parent couple who had contracted with Chanbua, David Farnell and Wendy Li, said that they would have asked for him to be aborted if his condition had been known sooner. In some accounts the Farnells wanted both Gammy and his non-Down twin sister, Pipah (whom they brought to Australia and are now raising); in other accounts Chanbua wanted to keep both children and the couple took the sister in fear that they would not be able to take both children

(Gammy was still in the hospital, and simultaneously there was political unrest in Thailand); and at some point it also emerged that the father had a record as a convicted child sex offender. By 2015 a story emerged that the Farnells had tried to lay claim to some of the money that individuals from around the world had donated to Chanbua to help her in raising Gammy, but ultimately a judge in Australia found them innocent of this charge and also affirmed their right to keep Pipah. See Kathy Marks, "Baby Gammy: Australian Father Who Abandoned Down Syndrome Surrogate Child Now Tries to Access Funds Donated for His Care," *Independent*, May 19, 2015, http://www.independent.co.uk/news /world/australasia/baby-gammy-australian-father-who-abandoned-down-syn drome-surrogate-child-now-tries-to-access-funds-10261916.html; and "Baby Gammy: Surrogacy Row Family Cleared of Abandoning Child with Down Syndrome in Thailand," *ABC*, April 14, 2016, http://www.abc.net.au/news /2016-04-14/baby-gammy-twin-must-remain-with-family-wa-court-rules /7326196.

    81.   Just to give one important example—already a decade old, but not yet widely discussed—involving the prospect of bearing a child with a disproportionate possibility of antisocial behavior, see Rosalie E. Ferner et al., "Guidelines for the Diagnosis and Management of Individuals with Neurofibromatosis 1," *Journal of Medical Genetics* 44, no. 2 (February 2007): 81–88. On the prospect of bearing a child with a heightened risk for schizophrenia, see Laura Hercher and Georgette Bruenner, "Living with a Child at Risk for Psychotic Illness: The Experience of Parents Coping with 22q11 Deletion Syndrome; An Exploratory Study," *American Journal of Medical Genetics* 146A (2008): 2355–60. For an overview of the state of the law on reproductive technologies across diverse nations within the EU, see Francesco Paolo Busardò et al., "The Evolution of Legislation in the Field of Medically Assisted Reproduction and Embryo Stem Cell Research in European Union Members," *BioMed Research International* 2014: 1–14. See also Bagińska, "Wrongful Birth"; Ruth Chadwick, "Can Genetics Counseling Avoid the Charge of Eugenics?," *Science in Context* 11 (1998): 471–80; Heather Paxson, "With or against Nature? IVF, Gender and Reproductive Agency in Athens, Greece," *Social Science and Medicine* 56 (2003): 1853–66; Hilmar H. Bijma et al., "Decision-Making after Ultrasound Diagnosis of Fetal Abnormality," *Reproductive Health Matters* 16, no. 31 (2008, supplement): 82–89; Reed Boland, "Second Trimester Abortion Laws Globally: Actuality, Trends and Recommendations," *Reproductive Health Matters* 18, no. 36 (November 2010): 67–89; Vicenzo Pavone and Flor Arias, "Beyond the Geneticization Thesis: The Political Economy of PGD/PGS in Spain," *Science,*

Technology, and Human Values 37, no. 3 (2012): 235–61; Ilana Löwy, "Zika and Microcephaly: Can We Learn from History?," *Physis: Revista de Saúde Coletiva* 26, no. 1 (2016).

82. For important reflections on changes in values already under way in the cultural paradigm shift from a desire for a right to privacy to more concern with a right to health, infertility treatment, and so on—but also with important suggestions for as yet untapped judicial approaches—see Judit Sándor, "Reproduction, Self, and State," *Social Research* 69, no. 1 (2002): 115–41. See also Rebecca J. Cook and Bernard M. Dickens, "Human Rights Dynamics of Abortion Law Reform," *Human Rights Quarterly* 25, no. 1 (February 2003): 1–59. And for a clarifying philosophical position on law and ethics in relation to new reproductive technologies that starts from the premise that we owe respect to "future persons" but that "embryos that do not have mothers willing to continue to assist their growth have no way of developing into persons," see Anja Karnein, *A Theory of Unborn Life: From Abortion to Genetic Manipulation* (New York: Oxford University Press, 2012) (quote from the back cover).

83. For a differentiated recent analysis—in the face of a rising tendency to castigate and shame women who choose to terminate a pregnancy after Trisomy 21 has been identified—of the wide variety of possible outcomes and the important point that the celebrated "poster people" are not necessarily typical, see the observations of French historian of medicine Ilana Löwy:

> A prenatal diagnosis of three 21 chromosomes does indicate that the child has a Down's syndrome, but not the severity of her/his intellectual and physical disability. Approximately 20 percent of individuals with Down's syndrome are defined as having a mild intellectual disability, 20 percent a profound one; the remaining 60 percent are somewhere between these extremes. Moreover, a significant proportion of Down's syndrome children have important health problems. Unsurprisingly, the "poster people" for Down's syndrome are among the high-performing ones. Usually they are also those who . . . were able to benefit from supportive family and from public aids, frequently secured thanks to their families' skill in navigating the official support system. To claim that the most successful cases should represent the future of every Down's child is, however, deeply misleading. It does not take into account important differences in public resources available for education and medical care of disabled people, important differences between socioeconomic status of mothers and families of such people and their psychological makeup, and above all important differences in the severity of impairment of individuals with Down's syndrome: some people with this condition can keep a regular job, but some never learn to speak. And

many will need a high level of parental support until the end of their parents'/ mother's life. As a rule, parents take care of their children when they are small, and not infrequently are helped by their children when they grow old. Parents/ mothers of intellectually disabled children know, however, that they will always stay on the giving end. It is not rare today for a woman in her seventies to have the sole responsibility of an adult child with Down's syndrome who, in addition, may suffer from an early onset of Alzheimer's disease, one of this condition's late effects. And, in today's economic and political climate, it may be difficult to tell a pregnant woman that she should count on an important increase in the level of public support for people with intellectual disabilities in the near future. A woman diagnosed with a Down's syndrome fetus nearly always faces a very painful personal decision, but also a risk that she be negatively judged for that decision. If she decided to maintain the pregnancy, she may be criticized for potential harm to her family and society. If she decides to interrupt it, she may be criticized for selfish behavior, absence of maternal virtues and a "eugenic" rejection of the diversity of human kind. . . . A book which—with the best possible intentions—minimizes the real-life problems of care of people with Down's syndrome, and implicitly condemns women who elect to terminate a pregnancy with a trisomic fetus, can make a very difficult situation of these women even more challenging. Ilana Löwy, review of David Wright, *Down's: The History of a Disability*, *New Genetics and Society*, March 2014, 229–31. See also the nuanced, balanced, and remarkable new book that is based on fieldwork in France (where termination of pregnancy is legal) and Brazil (where it is not): Löwy, *Tangled Diagnoses: Prenatal Testing, Women, and Risk* (Chicago: University of Chicago Press, 2018). Löwy here eloquently argues both for the extraordinary complexity of individuals' situated decision-making and for the imperative "to respect equally the decisions of women who feel unwilling or unable to risk raising a disabled child and those who refuse to terminate a pregnancy with a severely impaired fetus" (218).

84. For example, Austrian law professor Erwin Bernat in 2007—in the context of defending preimplantation diagnostics for in vitro fertilization— observed: "Whoever does not want a disabled child demonstrates just as little hostility to the disabled as the politician who advocates for a zero-tolerance-for-alcohol policy in operating an automobile—a policy that, in its tendency, surely contributes to lowering the number of severely disabled victims of traffic accidents." In general, he analogized: "A person can prefer the life-situation A to the life-situation B. He can decide to give up smoking in order to extend his

life expectancy. Likewise he can shift from being an agnostic to being a religiously affiliated person, because he 'starts to believe.' Does this person dis-
criminate against smokers and the religiously non-affiliated, if he ceases smoking and commits to a religion? The answer should be self-evident for everyone."
In sum, then—and although he was referring to Austrian, not German, law—
the point holds: "The constitutionally guaranteed prohibition against discrimination of the disabled . . . thus does not mean that it can be morally right to
evaluate *the* illness or *the* disability of human beings per se as positive. For if the
illness or the disability per se had a positive value then measures to heal or prevent illness or disability would be not only completely incomprehensible. We
would actually be required, should the occasion arise, to produce people with
disabilities. Nobody, however, can for good reasons want such a consequence
[*Eine solche Konsequenz kann aber niemand mit guten Gründen wollen*]" (Erwin
Bernat, "Pränatale Diagnostik und Präimplantationsdiagnostik auf dem Prüf-
stand des österreichischen Rechts," in *Rechtsfragen der Präimplantationsdiagnostik*, ed. Christian Dierks et al. [Berlin: Springer, 2007], 25–63, here 53). For
a more intimate and personal perspective, see the article by Gillian Relf, "'I
Wish I'd Aborted the Son I've Spent 47 Years Caring For': It's a Shocking
Admission—but Read On before You Judge," *Daily Mail*, October 22, 2014,
http://www.dailymail.co.uk/femail/article-2803834/I-wish-d-aborted-son-ve
-spent-47-years-caring-s-shocking-admission-read-judge.html#ixzz4ZHiy23n7.

85. Remarks by Jane Fisher of Antenatal Results and Choices, testifying in
the UK Parliamentary Inquiry into Abortion on the Grounds of Disability,
2013, https://web.archive.org/web/20130604174731/http://www.abortionand
disability.org/page3/page3.html. Among other things, in response to MPs'
questions, Fisher noted:

> I would add that I certainly don't think when they [parents] make that very
> difficult and very personal decision that they are making any kind of statement
> on those that are actually living with disability or impairment. This is a decision
> that's incredibly pertinent to their individual circumstances, and certainly out-
> side the scope of post-24 weeks we've supported many couples who have first-
> hand experience of living with disability because they carry genetic conditions
> or have family members or work with people with disabilities, and I know so
> well from my personal experience that they're certainly not wishing to devalue
> those living with disability when they make their own decision. . . . What parents
> know is that it is a spectrum. So there are many children that do very well and
> lead semi-independent lives. There are others that struggle much, much more
> with their level of learning disability and with associated health issues, and

there often is the crux of the difficulty for the parents involved, and it's why I think we've seen in these years where quite rightly the disability rights lobby have made huge inroads in encouraging a more inclusive society—we still see that the majority of people who are given a prenatal diagnosis of Down's Syndrome make the difficult decision to end the pregnancy. They certainly do not make that decision lightly. They do not say to themselves, "Oh, my baby is not perfect, I no longer want it." They agonize over that decision. But they look at their own family circumstances, what that diagnosis might mean to them, because they cannot have a definitive prognosis, an absolute prognosis, and the majority think, because we really don't feel able to cope and we don't want our child to cope with the worst case scenario, we feel we're going to end the pregnancy ourselves. . . . Yes. I mean, I can accept this on many levels, that people living with a particular disability will feel very uncomfortable with the fact that some women choose to end a pregnancy on the strength of that disability. But they're two separate situations. And I've certainly spoken to many people with disabilities who, yes, would articulate that philosophical difficulty with it, but would also say, "I cannot say that, by withholding that woman's right to make her reproductive choice, I am furthering my cause in any way." I don't think that those two things work, and I do genuinely believe, and I think again my experience has borne this out, that there's not a tension for me in having a society that is empathetic, that is inclusive and enabling women to make their own autonomous reproductive choices. Those things to me go hand in hand.

86. One of the few significant countervailing voices is that of the Swiss politician Luc Recordon, himself an individual with disabilities. Expressly responding—to a journalist in 2015—to those who would argue that the availability and use of preimplantation diagnostics send a problematic message about the lack of valuing of disabled lives and thus also puts pressures on parents to abort in cases of disability discovered prenatally, Recordon demurred and noted several points. One was that it so happened that there appeared to be greater support for individuals living with disabilities in countries in which preimplantation diagnostics were permitted, so that the causal or correlational argument simply did not have an empirical basis. Second, Recordon was appalled at the implication that "somehow a society needs 'enough' disabled people so that the already existing disabled ones don't feel yet more isolated. . . . That is as if one, instead of combating poverty, would double the number of the poor so that they don't feel so alone." After all, he continued, "One could also reverse the argument and say that the fewer disabled people there are, the easier it is to help them. There is more money—that's plain mathematics." And

when asked what he wished for, Recordon said frankly: "A society with as few disabled people as possible, of course. It isn't fun to be disabled." Interview with Recordon by Noëmi Landolt, "Möglichst wenig Behinderte? Mais bien sûr," *Die Wochenzeitung*, May 21, 2015, http://www.woz.ch/1521/praeimplanta tionsdiagnostik/moeglichst-wenig-behinderte-mais-bien-sur. Already in 2005, when he addressed his colleagues in the Swiss parliament as he defended the use of preimplantation diagnostics, Recordon had made a third point, one that touched on the very morality of giving birth to a child with disabilities:

> I do not pretend to deliver here an objective point of view. I am much too emotionally involved in what I am going to tell you to be able to do it. But I believe that a testimony should be given here. I was born with a very heavy congenital disability [Holt-Oram syndrome, contributing to a malformation in the limbs that in his case was accompanied by exceptionally severe pain]. I was very lucky, because I was surrounded by responsible and loving parents who committed themselves with determination so that I could develop my activities and my personality. But I think it was a rare chance, and I've seen a lot of people in hospitals, in rehabilitation centers in Switzerland and Germany in particular, in the United States too, who have not had this chance, nor have they had the moral strength to resist disability, which crushed them. I believe that when one reads, in the words of some organizations of the disabled, that to accept preimplantation diagnosis, is to despise disability and the disabled, I believe deeply that it is quite the opposite. That is because we need to recognize that it is something humanly very heavy and that it is a misfortune, a terrible calamity for the majority of those who experience it [*C'est parce qu'on doit reconnaître que c'est quelque chose d'humainement très lourd et que c'est une malchance, une infortune terrible pour la plupart de ceux qui la subissent*]. . . . I am a little annoyed to hear the question: Is there a right to a healthy child? . . . Can a human being have a right to the life of another human being, even if it is in the positive sense of giving life? I certainly hope not! To make a child, that is to take on a terrible responsibility; and not to do everything possible to ensure that this child has every prospect of good fortune, that is a terrible way of not assuming responsibility.

National Council, Switzerland, *Amtliches Bulletin—Bulletin Officiel*, June 16, 2005, 14. See also Recordon's subsequent reaffirmation of his position in Xavier Alonso, "Luc Recordon, l'homme qui aurait préféré ne pas naître," *24 Heures*, June 28, 2012, http://www.24heures.ch/suisse/Luc-Recordon-l-homme-qui -aurait-prefere-ne-pas-naitre/story/18416220.

87. To give just one example: 60 percent of adults with cognitive disabilities

in Europe still live with their parents. See Inclusion Europe, "Families and In-tellectual Disability in Europe" (2002), http://supporteddecisionmaking.org /sites/default/files/Families%20and%20Intellectual%20Disability%20in%20 Europe.pdf, as well as the organization's website, http://inclusion-europe.eu/; everywhere the creation of joyful communities and "assisted freedom" (Sigrid Graumann's term) requires indefatigable parental initiatives. For an example, see Cigdem Akyol, "Mein Kind ist eine Belastung" (interview with Morlind Tumler and Götz Aly about the life they have created for their daughter Kar-line), *Wiener Zeitung*, June 6, 2013, http://www.wienerzeitung.at/themen_ channel/literatur/autoren/552453_Mein-Kind-ist-eine-Belastung.html; and Tumler and Aly's testimonial: http://www.zukunftssicherung-ev.de/testimonials/; as well as Sigrid Graumann, *Assistierte Freiheit: Von einer Behindertenpolitik der Wohltätigkeit zu einer Politik der Menschenrechte* (Frankfurt am Main: Campus, 2011).

Chapter 2. Moral Reasoning in the Wake of Mass Murder

1. Eugen Stähle quoted in Ernst Klee, "Von Menschen und Tieren: Eine Kritik der praktischen Ethik," *Die Zeit*, June 30, 1969, 58.

2. Götz Aly, *Die Belasteten: "Euthanasie" 1939–1945—Eine Gesellschaftsge-schichte* (Frankfurt am Main: Fischer, 2013); Maike Rotzoll et al., eds., *Die nationalsozialistische "Euthanasie"-Aktion "T4" und ihre Opfer: Geschichte und ethische Konsequenzen für die Gegenwart* (Paderborn: Schöningh, 2010); Michael Burleigh, *Death and Deliverance: "Euthanasia" in Germany, c. 1900 to 1945* (New York: Cambridge University Press, 1995); Henry Friedlander, *The Origins of Nazi Genocide: From Euthanasia to the Final Solution* (Chapel Hill: Univer-sity of North Carolina Press, 1995); Ernst Klee, *"Euthanasie" im NS-Staat: Die "Vernichtung lebensunwerten Lebens"* (Frankfurt am Main: Fischer, 1983).

3. On the killings of psychiatric patients in Poland, see Klee, *"Euthanasie" im NS-Staat*, 95–98; and Walter Grode, "Deutsche 'Euthanasie'-Politik in Polen während des Zweiten Weltkriegs," *Psychologie und Gesellschaftskritik* 16, no. 2 (1992): 5–13. See also the interview with Friedrich Leidinger, *Schatten-blick*, February 7, 2014, http://www.schattenblick.de/infopool/pannwitz/report /ppri0019.html.

4. The best recent analysis of the intricate mutual imbrication of the murder of the disabled and the murder of European Jewry—with special atten-tion not least to the role of erstwhile low-status asylum custodians as killers in

the Operation Reinhard camps—is Sara Berger, *Experten der Vernichtung: Das T4-Reinhardt-Netzwerk in den Lagern Belzec, Sobibor und Treblinka* (Hamburg: Hamburger Edition, 2013).

5. Helga Kuhse and Peter Singer, *Should the Baby Live? The Problem of Handicapped Infants* (New York: Oxford University Press, 1985), v.

6. For an overview of the unfolding events, see Martina Ahmann, *Was bleibt vom menschlichen Leben unantastbar? Kritische Analyse der Rezeption des praktisch-ethischen Entwurfs von Peter Singer aus praktisch-theologischer Perspektive* (Münster: LIT, 2001), 36–40.

7. *Der Spiegel* in 1984 cited as an estimation of the number of cases annually in West Germany of newborns permitted to die by withholding of care—among them cases of spina bifida, anencephaly, and Down syndrome without a functioning gastrointestinal system—at approximately twelve hundred.

8. Portions of the interview were reported fairly immediately in Hans Schuh, "Lässt sich Euthanasie ethisch begründen?," *Die Zeit*, June 16, 1989; a retrospective account of the significance of the Saarbrücken interview is provided by Georg Meggle, "Schwierigkeiten der Medien mit der Philosophie," *Telepolis*, March 22, 2005, http://www.heise.de/tp/artikel/19/19722/1.html. For criticisms of Meggle, see Oliver Tolmein, *Wann ist der Mensch ein Mensch?* (Munich: Hanser, 1993); and Henryk Broder, "The Nutsy Professor," *Die Achse des Guten*, July 2005, http://www.achgut.de/dadgd/view_article.php?aid=903.

9. Schuh, "Lässt sich Euthanasie ethisch begründen?" The further headlines came from, respectively, the *Dortmunder Rundschau* (May 24, 1989), the *Ruhr-Nachrichten* (June 10, 1989), and the *Westdeutsche Allgemeine Zeitung* (June 24, 1989). See Christoph Anstötz, "Peter Singer und die Pädagogik für Behinderte: Der Beginn der Singer-Affäre," *Analyse & Kritik* 12 (1990): 131–48, here 137. In addition, critics of Singer urged federal and regional government ministers to investigate university curricula where Singer's works were being read and/or called on faculty who had defended him formally to distance themselves from his views.

10. Quoted in Reinhard Merkel, "Der Streit um Leben und Tod," *Die Zeit*, June 23, 1989.

11. Hans Jonas from the United States and Richard Hare in the UK; see also the statement of the Aristotelian Society of Great Britain, July 16, 1989, reprinted in Christoph Anstötz, Rainer Hegselmann, and Hartmut Kliemt, eds., *Peter Singer in Deutschland: Zur Gefährdung der Diskussionsfreiheit in der*

Wissenschaft; eine kommentierte Dokumentation (Frankfurt am Main: Peter Lang, 1995), 11.

12. Peter Singer, "On Being Silenced in Germany," *New York Review of Books*, August 15, 1991, 36–42.

13. Anstötz, Hegselmann, and Kliemt, *Peter Singer in Deutschland*; for more on freedom of speech, see also the publisher's description of the book, available at https://www.amazon.com/Peter-Singer-Deutschland-Diskussions freiheit-Dokumentation/dp/3631480148.

14. See Merkel, "Der Streit." For more examples of discussions of "taboos" and "tabooizing," see the conversation in "Exzess der Vernunft"; Meggle, "Bemerkungen"; and Cornelia Filter, "Das Affentheater," *Emma*, March/April 1994, 68–73.

15. Anstötz, "Peter Singer," 136, 141. On the connections between disability activism and HIV/AIDS-related activism (and Singer), see also Horst Ladenberger, "Ein Menschenbild von Behinderten gab's schon immer: Versuch einer Standortbestimmung," in Deutsche AIDS-Hilfe e.V., "'Was heißt'n hier behindert?': Berührungspunkte von Behindertenbewegung und AIDS-Hilfen" (1997), https://www.aidshilfe.de/sites/default/files/documents/Was %20hei%C3%9Ftn%20hier%20behindert%20Dokumentation%201997.pdf. Other groups among the protesters were from such organizations as the Association of Disabled and Nondisabled Students, the (national) Alliance of Disabled and Cripple Initiatives, the student governments of the Universities of Dortmund and Cologne, and the German Society for Social Psychiatry.

16. Christoph Anstötz, Rainer Hegselmann, and Hartmut Kliemt, "Einführung," in Anstötz, Hegselmann, and Kliemt, *Peter Singer in Deutschland*, 6–8. The Nazi text cited is Raymund Schmidt, "Das Judentum in der Philosophie," in *Handbuch der Judenfrage*, 38th ed. (1935).

17. Meggle, "Bemerkungen," 33, also quoted in Martin Blumentritt, "Das 'bioethische' Netzwerk" (last updated April 7, 1998, but written originally in 1990 or 1991), http://www.comlink.de/cl-hh/m.blumentritt/agr218s.htm. In addition to tracking the defenders of Singer and the various ways they dug in their heels and successfully advanced, both in Germany and internationally, the promulgation of Singer's and related bioethical theories, this Blumentritt essay also summarizes the various evolving self-distancings and subsequent self-criticisms of some initial defenders of Singer's right to promote his ideas.

18. Peter Singer, "Bioethics and Academic Freedom," *Bioethics* 4, no. 1 (January 1990): 33–44, here 42.

19. "Wider den tödlichen philosophischen Liberalismus," *Die Randschau* 5, no. 1 (January–April 1990): 24.

20. For a critic of Singer who refers to it as "the Singer affair," see the philosopher Martin Blumentritt, "Von Singer zu Hitler," April 7, 1998, http://www.comlink.de/cl-hh/m.blumentritt/agr317s.htm; for a defender/promoter of Singer who did so already in 1990, see Anstötz, "Peter Singer"; for another defender who did so, see Meggle, "Schwierigkeiten der Medien."

21. Klee, *"Euthanasie" im NS-Staat*. In this book, Klee had a longer version of the Stähle quote, made in response to a Protestant churchman named Sautter on December 4, 1940 (the records came from the trial of the doctors involved at the killing center of Grafeneck): "Where the will of God really counts and is followed through on, namely, in free nature, there is no mercy for the weak and sick. . . . No ill rabbit can hold on longer than a few days: it becomes the certain plunder of its enemies and is thus saved from its suffering; that is why rabbits are always a 100 percent healthy society. . . . The fifth commandment, Thou shalt not kill, that is not divine law, that's a Jewish invention" (16). The reference to animals as ruthless toward the weak and in this way a model for humans was another major Nazi theme. See, for instance, the film *Alles Leben ist Kampf* (All life is struggle, 1937), shown to NSDAP Party members for training purposes. It begins by expounding on fights for survival among animals but concludes with images of visibly disabled individuals with the clear implication, here also given a pseudoreligious gloss (implying that Nature's way of preferring the strong is God's will as well), that they should be done away with. See U.S. Holocaust Memorial Museum, "The Disabled and Heredity," https://www.ushmm.org/online/film/display/detail.php?file_num=2553. Klee had a distinctive style, combining a wealth of at once emotionally shattering and highly evocative empirical details, frequently tellingly juxtaposed, with a deadpan searing sarcasm of tone. He was often resented by medical professionals—though they might grant that he had "a good nose" (*eine gute Nase*) for incriminating factoids—and by academic historians. Already in 1974 and 1980 Klee had published journalistic exposés of the conditions in institutions for the disabled, and he was one of the co-organizers, with Gusti Steiner, of the first militant public demonstration for disability rights, in 1975—a wheelchair-induced traffic jam in the city center of Frankfurt am Main at rush hour to demand curb cutting and the removal of other barriers to public access for individuals with disabilities. The pioneering English-language study by Henry Friedlander of the murders of the disabled, *The Origins of the Nazi Genocide*

(1995), was in part made possible by materials that Klee gave him—sources that archivists and other guardians of implicating data had been keeping from seeing the light of day.

22. Henry Friedlander, remarks at the "Lessons and Legacies" conference of the Holocaust Educational Foundation in Evanston, IL, in 2008.

23. Carol Poore, *Disability in Twentieth-Century German Culture* (Ann Arbor: University of Michigan Press, 2007); Stefanie Westermann et al., eds., *NS-"Euthanasie" und Erinnerung: Vergangenheitsaufarbeitung—Gedenkformen—Betroffenenperspektiven* (Münster: LIT, 2011); Sascha Topp et al., eds., *Silence, Scapegoats, Self-Reflection: The Shadow of Nazi Medical Crimes on Medicine and Bioethics* (Göttingen: V&R unipress, 2014); Ralf Forsbach, "Die öffentliche Diskussion der NS-Medizinverbrechen," in *NS-Medizin und Öffentlichkeit: Formen der Aufarbeitung nach 1945*, ed. Stephan Braese et al. (Frankfurt am Main: Campus, 2016).

24. Ernst Klee, *Was sie taten—was sie wurden: Ärzte, Juristen und andere Beteiligte am Kranken- oder Judenmord* (Frankfurt am Main: Fischer, 1986); Peter Hayes, *Why? Explaining the Holocaust* (New York: W. W. Norton, 2017).

25. Already in 1980, the historian, psychiatrist, and psychiatric reform activist Klaus Dörner had sought to harness the—unexpectedly emotionally involved and empathetic—reaction of millions of non-Jewish German viewers to the plight of persecuted and murdered Jews, presented in the broadcast on German television in 1979 of the American miniseries *Holocaust*, specifically for the purpose of engaging Germans in concern about conditions for psychiatric patients and individuals with disabilities in postwar Germany. Dörner's manifesto-cum-document collection was "dedicated to the psychiatrically, mentally, and physically disabled citizens killed in the 'Third Reich' and their families." In its beginning pages it invited readers to "think about the comparison 1940–1979," and it was titled in such a way as to invoke the American miniseries and to emphasize the reciprocal relationships between past and present and between the Holocaust and the T4 and other killings of the disabled: *Der Krieg gegen die psychisch Kranken: Nach "Holocaust" Erinnern, Trauern, Begegnen* (The war against the psychiatrically ill: After "Holocaust" remembering, mourning, encountering). In an extended explanation of the dedication, Dörner—who at that point thought the total of murdered disabled was 120,000 (a sign of just how unadvanced the historiography on these murders was at that historical moment)—remarked both that these victims had, "of all victims of the NS-regime, died the loneliest and most unwitnessed of deaths"

and that it was especially important also to demonstrate solidarity with their families, for "of all those left behind by NS-victims these are the ones who have been most abandoned by us in their suffering. This persecuted group was the only one not recognized by us as persecuted and was denied compensation." And "because we who are involved in the work of psychiatry, we who should have been responsible, defending against guilt and fear, blind and silent, we have abandoned their families in their isolation." Dörner expressly called on readers to meditate on the fact that on September 1, 1939, not solely "the extermination war to the outside" (the attack on Poland) had begun but also "the extermination war to the inside," as Hitler had deliberately backdated his order to kill disabled children to that same date. Additionally, he asked readers to ponder "to what extent and how this war to the inside is still continuing today." The self-referential self-critical "us" and "we" was deliberate. The book opened with strikingly self-lacerating testimonials from psychiatric care workers about their sense of shame at the persistence of inadequate care for the disabled and for psychiatric patients; included a memorandum from the German Society for Social Psychiatry from September 1, 1979, demanding psychiatric reform, as well as (and despite official lack of interest from the Federal Ministry of Health) numerous politicians' and other public figures' positive responses to the memorandum; and also reprinted scores of primary sources from the Nazi era from both perpetrators and objectors documenting the killings.

26. Note the assertive-informative subtitle to a headline in an article by Ernst Klee in 1990: Ernst Klee, "Der alltägliche Massenmord: Die 'Euthanasie'-Aktion war der Probelauf für den Judenmord—der Kreis der Opfer wurde bis Kriegsende immer mehr erweitert" (The everyday mass murder: The "euthanasia" action was the trial run for the Judeocide—the circle of victims was expanded ever more until the end of the war), *Die Zeit*, March 23, 1990.

27. As he went on to elaborate this point, Singer showed considerable familiarity with the facts of Nazi euthanasia, yet simultaneously he insisted again on the categorical distinctions he saw between the two mass crimes (a sign once more, as if another were needed, of the complex relationship between facts and interpretation): "In the case of Nazism, it was the racist attitude towards 'non-Aryans'—the attitude that they are sub-human and a danger to the purity of the Volk—that made the holocaust possible. Nor was the so-called 'euthanasia' programme anything like the kind of euthanasia that could be defended on ethical grounds—as can be seen from the fact that the Nazis kept their operations completely secret, deceived relatives about the cause of death of those killed, and exempted from the programme certain privileged classes, such

as veterans of the armed services, or relatives of the euthanasia staff." And "'Doing away with useless mouths'—a phrase used by those in charge—gives a better idea of the objectives of the programme than 'mercy-killing.' Both racial origin and ability to work were among the factors considered in the selection of patients to be killed. There is no analogy between this and the proposals of those seeking to legalize euthanasia today" (Peter Singer, *Practical Ethics* [New York: Cambridge University Press, 1979], 154–56).

28.   An activist organization geared to achieving recognition and reparations for victims of involuntary sterilization and for family members of individuals murdered in the Nazi killings of the psychiatrically ill and cognitively handicapped linked the two matters in its very name. The Bund der "Euthanasie"-Geschädigten und Zwangssterilisierten e.V. (Federation of "Euthanasia"-Harmed and Coercively Sterilized People) was launched in the town of Detmold in 1987. See Margaret Hamm, "Zwangssterilisierte und 'Euthanasie'-Geschädigte: Ihre Stigmatisierung in Familie und Gesellschaft," in Rotzoll et al., *Die nationalsozialistische "Euthanasie"-Aktion "T4,"* 358–63; and Sascha Topp, *Geschichte als Argument in der Nachkriegsmedizin: Formen der Vergegenwärtigung der nationalsozialistischen Euthanasie zwischen Politisierung und Historiographie* (Göttingen: Vandenhoeck & Ruprecht, 2013). Previously, these individuals had not had any ability to connect with each other but had each struggled in isolation both in negotiations with the authorities and in coming to terms with their own traumas. Also in 1987 the Green Party formally took up their cause. In April of that year the party brought a proposal to the Bundestag to demand compensation for victims of involuntary sterilizations. The Bundestag rejected the proposal, claiming constitutional impediments, as it had never been clearly decided whether the law of 1933 had been rescinded by the military occupiers in 1945 or whether it had remained in force. As a separate matter, the Bundestag considered whether the Basic Law promulgated in 1949—West Germany's equivalent of a constitution—had obviated any need to repudiate Nazi legislation, since it simply superseded it. These irresolvable niceties kept anything from happening, and, in addition, legislators admitted that they were worried about the budgetary implications should victims of involuntary sterilization be legally permitted to lay claim to compensation. The ensuing unsatisfactory compromise led to the Bundestag in 1988 repudiating the judicial decisions made on the basis of the 1933 law but not the law itself. Nonetheless, this decision of 1988 did mark a shift in that the involuntary sterilizations began to be seen as the injustices they had been. (Not until 2007 did the Bundestag repudiate the law itself.)

29. Postwar experts strenuously contested the idea that the sterilizations could be understood under the rubric of racism. To give just one example: Hans Nachtsheim, a physician who had in the 1940s—historians found— "conducted experiments on children who had epilepsy and on organs taken from prisoners murdered at Auschwitz," was on the committee appointed by the West German government in the early 1960s to decide whether victims of coerced sterilizations should receive compensation. In a medical journal, Nachtsheim wrote in 1962 in an essay titled "About the Law for the Prevention of Hereditarily Diseased Offspring of 1933 from the Present-Day Perspective" that "we must at long last stop throwing the Law for the Prevention of Hereditarily Diseased Offspring into the same pot with National Socialist racial laws." Among his reasons was that "thereby a future law on hereditary diseases, which must come, is once again preemptively discredited." See Svea Luise Hermann and Kathrin Braun, "Das Gesetz, das nicht aufhebbar ist: Vom Umgang mit den Opfern der NS-Zwangssterilisation in der Bundesrepublik," *Kritische Justiz* 43, no. 3 (2010): 338–52, here 344. The Nachtsheim essay they are quoting is Hans Nachtsheim, "Das Gesetz zur Verhütung erbkranken Nachwuchses aus dem Jahre 1933 aus heutiger Sicht," *Ärztliche Mitteilungen* 59 (1962): 1640 ff. Nachtsheim had initially conducted experiments on rabbits to see whether reduced air pressure through loss of oxygen could trigger seizures; in 1943 he transferred these experiments to six epileptic children from the Brandenburg facility. In the case of organs from Auschwitz, it was the eyes of victims that he worked with. Another theme that concerned him in the early 1960s were the birth defects caused by thalidomide; Nachtsheim downplayed the causative effect of the pharmaceutical and instead emphasized purported hereditary vulnerability of the pregnant women affected.

30. Uwe Kaminsky, "Eugenik und 'Euthanasie' nach 1945: Historiografie und Debatten am Beispiel der Evangelischen Kirche," in Rotzoll et al., *Die nationalsozialistische "Euthanasie"-Aktion "T4."*

31. A number of historiographical interventions changed the conversations considerably. By 1983 the German historian Gisela Bock had published a pioneering article in English titled "Racism and Sexism in Nazi Germany: Motherhood, Compulsory Sterilization, and the State"; by 1986 her comprehensive German-language book on the compulsory sterilizations in the Third Reich had been published, with the subtitle clearly announcing that the topic concerned "racial politics and gender politics." In 1985 a group of young New Left–linked historians around Götz Aly and Karl Heinz Roth began publishing a series titled *Beiträge zur nationalsozialistischen Gesundheits- und Sozialpolitik*

(Contributions to National Socialist health politics and social politics), which took on numerous topics relating to eugenics and euthanasia and saw itself as documenting how Nazi social policy needed to be understood as racial policy. All of this was happening in a political climate in which it was increasingly common to call for both research and activism on behalf of what were, by that point, often called the "forgotten victims" (the "asocials," the victims of sterilization, the handicapped and psychiatric patients, Roma and Sinti, and male homosexuals). And while this was a somewhat infelicitous description, since the use of the term could, and did at times, imply that somehow Jews had gotten "too much" attention (an implication that failed to acknowledge just how much antisemitic resentment had accompanied that attention), it was apt insofar as it accurately described groups whose efforts to be included as persecutees had often been rebuffed. Furthermore, in the midst of all these developments, historians began not just to study either the sterilizations or the murders but to think them together. In 1987 the historian Hans-Walter Schmuhl published an influential book whose title (in English, *Racial Hygiene, National Socialism, Euthanasia*) indicated its commitment to seeing both the sterilizations and the murders of the disabled within a "racial hygiene" framework and whose subtitle (in English, *From the Prevention to the Extermination of "Life Unworthy of Life," 1890–1945*) made explicit Schmuhl's conviction—shared by many—that the coercive sterilizations had indeed been a stepping-stone on the path to mass murder. Were the sterilizations in fact such a stepping-stone? Some scholars begged to differ, insisting on keeping eugenics and euthanasia analytically distinct. But in the back-and-forth of scholarly arguments of the time, their position was seen primarily as an effort to keep some historical variants of eugenics (e.g., Weimar-era socialist eugenics) free of the taint of the euthanasia murders. And in 1991 a notable anthology edited by the historian Norbert Frei, *Medizin und Gesundheitspolitik in der NS-Zeit* (Medicine and health policy in the NS era), sponsored by the prestigious Institut für Zeitgeschichte (Institute for Contemporary History) in Munich, brought the new scholarly questions and answers—including the work of Schmuhl—to an even broader audience. Also by 1991 the crucial critical English-language synthesis of the accumulating research was published, a British-German coproduction: Michael Burleigh and Wolfgang Wippermann's *The Racial State: Germany 1933–1945* (Cambridge: Cambridge University Press, 1991). A few years later still, Burleigh would also publish *Death and Deliverance*, his major study of the disability murders. Finally, by the mid-1990s all this work began to be absorbed into the field of Holocaust studies as well—a field not just newly expanding due to the opening of Eastern

Bloc and Soviet archives in the wake of the collapse of Communism but also acquiring broader standing as the US Holocaust Memorial Museum opened its doors in 1993. A signal development was the publication of the American historian Henry Friedlander's *The Origins of the Nazi Genocide* in 1995. Friedlander recounted, in the book's opening pages, that it had been "by the mid-1980s" that "my reading of the documents had convinced me that the euthanasia program had been intimately connected to Nazi genocide." Nonetheless, "I still thought of euthanasia as only a prologue to genocide." It was because of work done by German scholars—including archival documents given him by Ernst Klee, but also such work as that of the geneticist and historian Benno Müller-Hill on scientists' involvement in the crimes of Nazism (profiled, as Schmuhl had been, in the Frei anthology)—that Friedlander began to see that "euthanasia was not simply a prologue but the first chapter of Nazi genocide." Friedlander's new conception was that it had been a mistake that prior "historians have categorized the Nazis' murder of the European Jews as totally different from the murder of other groups." For Friedlander, the disabled, like the Jews, had been targeted "because they belonged to a biologically defined group. . . . Jews were not the only biologically selected target." Friedlander, *The Origins of the Nazi Genocide*, xii. Friedlander thus demonstratively began with a chapter on the sterilization program both for the disabled and for Roma and Sinti (whom Friedlander also believed should be counted as victims of the Holocaust). For as one reviewer appreciatively noted (it was Benno Müller-Hill, writing in the respected *Frankfurter Allgemeine Zeitung*)—and here one can see the evolving scholarly consensus taking shape—"from depriving the disabled of their civil rights [as exemplified by the sterilizations] a direct path led to their murder in the 'euthanasia project'" ("Rezension: Sachbuch Massenmord und Raubmord," *Frankfurter Allgemeine Zeitung*, February 16, 1998, http://www.faz.net/aktuell/feuilleton/politik/rezension-sachbuch-massenmord-und-raubmord-11308437.html). Once more, the framework that connected the sterilizations and the murders was confirmed.

32. Burleigh and Wippermann, *The Racial State*, 3–4.

33. See the summary in Meggle, "Bemerkungen."

34. The best explanations of the recent findings, based on sources discovered after the fall of the Wall and worked through carefully between 1995 and 2006 (noting the smaller-than-assumed overlap between the categories of victims who were sterilized and those who were murdered, finding that economic factors and capacity to work played a larger role in escaping the gas chambers than Nazi racial theories and hereditary-biological factors), and of why these

explanations have been perceived as threatening although they need not be, can be found in Maike Rotzoll et al., "Die nationalsozialistische 'Euthanasie'-Aktion 'T4' und ihre Opfer: Von den historischen Bedingungen bis zu den Konsequenzen für die Ethik der Gegenwart; Eine Einführung," and Marion Hulverscheidt, "Zusammenfassung der Podiumsdiskussion: 'Die Selektion: Neue Erkenntnisse?,'" both in Rotzoll et al., *Die nationalsozialistische "Euthanasie"-Aktion "T4"*; and Herwig Czech, "Nazi Medical Crimes, Eugenics, and the Limits of the Racial State Paradigm," in *Beyond the Racial State*, ed. Devin O. Pendas et al. (New York: Cambridge University Press, 2017). Czech in what might be termed a turn to a "bioeconomic" understanding of racism—notes that "eugenic" sterilizations and "euthanasia" murders were *both* often based on social diagnostics and cost-effectiveness calculations. Notably, and coinciding in time with the 2006 conference in Heidelberg, where many of the most important new findings were first controversially presented, the Bundestag—finally, after decades of sustained activism to achieve this end—moved to repudiate not just the coercive sterilization verdicts (which it had done in 1988) but also the 1933 Law for the Prevention of Hereditarily Diseased Offspring itself. In the application for this formal repudiation—which is deeply moving to read—the historiographical achievements of the 1980s, especially the connecting of "eugenic" sterilizations as a "precursor step" to the "euthanasia" mass murders, remain fully in evidence. The repudiation was co-sponsored by Christian Democrats / Christian Social Union and Social Democrats. See Antrag der Abgeordneten Dr. Jürgen Gehb et al., "Ächtung des Gesetzes zur Verhütung erbkranken Nachwuchses vom 14. Juli 1933," Deutscher Bundestag, Drucksache 16/3811, http://dip21.bundestag.de/dip21/btd /16/038/1603811.pdf.

35. A flood of articles in national and regional newspapers, numerous television programs, and multiple books written for popular audiences testified to the sudden presence in the public sphere of the radical disability rights movement's perspectives. An early recognition of having arrived can be seen in Franz Christoph, "(K)ein Diskurs über 'lebensunwertes Leben'!," *Der Spiegel*, June 5, 1989, 240-42. Singer noted the phenomenon as well, though he also took note of the pleasing irony that the controversy had revived flagging sales of *Practical Ethics*: "The book sold more copies in the year after June 1989 than it had in all the five years it had previously been available in Germany" ("On Being Silenced," 40).

36. Aviad Raz, "'Important to Test, Important to Support': Attitudes toward Disability Rights and Prenatal Diagnosis among Leaders of Support

Groups for Genetic Disorders in Israel," *Social Science and Medicine* 59 (2004): 1857–66, here 1857. The view from outside does suggest that the German situation need not have turned out the way it did. Looking in the first decades of the twenty-first century at the situation that has evolved in Germany, Israeli genetic counselors have criticized the official German position of moral outrage directed specifically and solely at abortion on grounds of fetal disability as "'absurd,' 'impractical,' 'high-minded' and even 'bullshit'" (Yael Hashiloni-Dolev and Aviad E. Raz, "Between Social Hypocrisy and Social Responsibility: Professional Views of Eugenics, Disability and Repro-genetics in Germany and Israel," *New Genetics and Society* 29, no. 1 [2010]: 87–102, here 94). In Israel, disability rights groups have tended toward the "two-fold view of disability," one in which they respond "in favor of prenatal genetic testing as well as selective abortion, while at the same time expressing their commitment for already-born individuals." The general idea is that it is "important to test" and also "important to support"—and these two are seen not as contradictory but rather as mutually complementary (Raz, "'Important to Test'"). Indeed, increasingly also ultra-Orthodox communities, traditionally uncomfortable with prenatal diagnostics and opposed to abortion, are developing new rabbinic leadership in "koshering" both—always with a view to the specific ability, or absence thereof, of a particular family lovingly to handle the birth and care of a disabled child. See Tsipy Ivry, "The Predicaments of Koshering Prenatal Diagnosis and the Rise of a New Rabbinic Leadership," *Ethnologie Française* 45, no. 2 (2015): 281–92. Interestingly, moreover, it is also Israeli social scientists who have managed to elicit from German genetic counselors what they have summarized as the "hypocrisy" and "non-honest discussion" in the German situation. Anonymously, German genetic counselors have commented on "a group dynamics that blocks moral thought" and "leads . . . to moral hypocrisy." "I personally," one genetic counselor admitted, "feel that it is very difficult for me to state a non-official opinion that does not have to do with high moral principles but with the difficult reality of human life. . . . Everybody has to support the official opinion, and hide what he really thinks, or he may be strongly criticized." Or, as another put it: "I believe the parents of the disabled are not so different in Germany than elsewhere, privately they don't want another sick child, but they don't dare argue in public. The ones who argue are the talented and the smart among the handicapped, those with political power. The thing is that only they are heard, not the really miserable ones." Moreover, this person continued: "I do believe society is richer with the disabled. But who has to pay the check for the education society needs? At the end of the day it is the disabled themselves, and their families, who have to deal with all the problems, and I don't see it as their duty to live in order to educate us. I don't think my patients should suffer for me to have an educational humanistic experience" (quoted in Hashiloni-Dolev and Raz, "Between Social Hypocrisy and Social Responsibility," 93–95).

37. For example, see "Sterbehilfe—der Tod als Freund," *Der Spiegel*, February 10, 1975, 36–60.

38. For a powerful ethical counterargument that passive letting-die by withdrawing care was the better course, see the interview with the New School philosopher Hans Jonas: "Mitleid allein begründet keine Ethik," *Die Zeit*, August 25, 1989.

39. Peter Singer, "Sanctity of Life or Quality of Life?," *Pediatrics* 72, no. 1 (July 1983).

40. Singer, *Practical Ethics* (1979), 97. Also in his 2011 revision, Singer persisted with this point: "We should reject the doctrine that killing a member of our species is always more significant than killing a member of another species. Some members of other species are persons; some members of our own species are not. . . . So it seems that killing a chimpanzee is, other things being equal, worse than the killing of a human being who, because of a profound intellectual disability, is not and never can be a person" (*Practical Ethics* [Cambridge: Cambridge University Press, 2011], 101).

41. Klee, "Von Menschen und Tieren."

42. Christoph had contracted polio one year after birth and had been subjected to multiple, severely painful operations in the years since. After the incident with Carstens, Christoph complained that he was being condescended to as a disabled person by having been permitted to stay in the events hall after the act rather than being arrested on the spot as a nondisabled person would have been. In 1983 he had protested the peace movement's demeaning representations of disability as among the damages caused by war. And in 1987 he had occupied the foyer of *Der Spiegel*'s main offices because in its coverage of a rise in reported cases of Down syndrome in the wake of the Chernobyl nuclear reactor disaster a year earlier the newsmagazine had remarked in passing, infelicitously, that Down was "the most common, just barely still compatible with life, malformation" (die häufigste, gerade noch mit dem Leben zu vereinbarende Missbildung) ("Bedrohung der Kinder," *Der Spiegel*, April 13, 1987, 237). Now, in June 1989, in the heat of the arguments over Singer, the newsmagazine was giving him an open forum.

43. Christoph, "(K)ein Diskurs."

44. Christoph, "(K)ein Diskurs."

45. Singer, "On Being Silenced."

46. Singer, *Practical Ethics* (1979), 126, 131.

47. Before Singer, the philosopher Michael Tooley had similarly used the acceptability of abortion as a reason to raise questions about the possible acceptability of infanticide—and Tooley had already mockingly noted that one of a "liberal's" biggest philosophical difficulties involved the problem of "specifying a cutoff point" at which a being had a right to life (Tooley clearly disagreed that viability or birth should be the place to draw the line between permissible and impermissible killing). Although less insistent on the issue of animal rights and less repetitively focused on the killing of disabled children, both themes were already in Tooley as well, as was "the question of what makes something a person." In Tooley's curiously passive phrasing, "Most people would prefer to raise children who do not suffer from gross deformities or from severe physical, emotional, or intellectual handicaps. If it could be shown that there is no moral objection to infanticide the happiness of society could be significantly and justifiably increased" ("Abortion and Infanticide," *Philosophy and Public Affairs* 2, no. 1 [Autumn 1972]: 37–65, here 38–39).

48. For 1946, see Dagmar Herzog, *Sex after Fascism: Memory and Morality in Twentieth-Century Germany* (Princeton, NJ: Princeton University Press, 2005), 75–76. "Thousandfold killing" is in the press release of the Bewegung für das Leben, December 12, 1984, reprinted in *Die Randschau* 1, no. 3 (August/September 1986): 13.

49. Helen Keller Kreis in the press release, p. 13; Ulrich Ochs (Waldbronn near Karlsruhe), letter to Federal President Philipp Jenninger, no date (presumably 1985), also reprinted in *Die Randschau* 1, no. 3 (August/September 1986): 13. The letter was also sent to all major newspapers. The letter included the statement: "I can see no difference between the racial ideology of Hitler and the 'criminal code reform' of the abortion Paragraph 218. Through the so-called 'eugenic indication' the disabled are sent a second time to a 'modern Auschwitz.'" Ochs again came into public in 2016 with the touching and happy news that he (as it turns out, of short stature) had fallen in love with and married a woman (with cerebral palsy) and that she, despite her own disability, had borne their little boy (who, in their view luckily, had inherited his condition). See Nina Job, "Handicap-Familie: 'Was Besseres konnte uns nicht passieren,'" *Abendzeitung* (Munich), January 18, 2016, http://www.abendzeitung-muenchen .de/inhalt.mutter-und-vater-mit-behinderung-handicap-familie-was-besseres -konnte-uns-nicht-passieren.ff70e512-e3fa-4998-a0de-a50039dab9d3.html.

50. Both the Catholic antiabortion flyer and the response of the Action Group are reprinted in Franz Christoph, *Krüppelschläge: Gegen die Gewalt der Menschlichkeit* (Reinbek: Rowohlt, 1983), 34. For a retrospective, highly positive view of 1981—seen globally—see Monika Baar's "Rethinking Disability" project at the University of Leiden, http://www.universiteitleiden.nl/en/re search/research-projects/humanities/rethinking-disability-the-global-impact -of-the-international-year-of-disabled-persons-1981-in-historical-perspective.

51. Krüppelgruppe Bremen, September 20, 1985, responding to a letter from Michael Drayss of the Bewegung für das Leben, reprinted in *Die Rand-schau* 1, no. 3 (August/September 1986): 14

52. The book included as well the stories of disabled women testifying to the horrendously condescending experiences they had endured when they sought abortions for unwanted pregnancies—as doctors, far from objecting to the women's desire for terminations, eagerly provided them (often in conjunction with recommending sterilizations), specifically because the doctors did not believe that disabled women should reproduce at all. For example, one woman who used a wheelchair wrote about her reasons for choosing abortion. (She had forgotten to take her birth control pill one day; her boyfriend had just started his university studies and was worried that he might, if they chose parenthood, have to break these off again.) The woman recounted how distressing it had been that while she was quickly granted the right to an abortion, she was told that it would be much easier to get it on the grounds of a maternal health indication (i.e., an implicit judgment on her lack of fitness, because of her disability, to raise a child) than on the grounds of a social indication (which was actually more accurate to her situation). She was also dismayed that one of the doctors whose approval she needed both treated her contemptuously and—again in view of her disability—suggested that she go ahead and take the opportunity of the abortion to get sterilized as well; she was terrified that she would be sterilized against her will, and she never knew whether in fact during the dilation and curettage procedure more of her had been "scraped out" than was necessary. Another contributor to the volume told a similar story of having had doctors prefer—with reference to the disability—the medical maternal health indication to the social indication, even though once again the social indication was more suited to the circumstances. Yet a further contributor encouraged all women with disabilities, should they need an abortion, to demand the social indication in order to be seen as female "and no longer as a neuter." See Silke Boll et al., *Geschlecht: Behindert, besonderes Merkmal: Frau—ein Buch von behinderten Frauen* (Munich: AG SPAK, 1985). But the volume additionally

included an essay, "We Cripple Women and Paragraph 218," that directly tackled the "eugenic indication," labeling it, unequivocally, as in itself "a discrimination of cripples." Acknowledging that it might seem "contradictory" to be, on the one hand, battling for women's self-determination and rejecting all scrutiny of women's reasons for seeking abortions while, on the other, labeling the eugenic indication in itself discriminatory, expressing awareness that "up until now only the reactionary, clerical crowd has taken up the embryopathic indication . . . and we need to be very careful not to give these people grist for their mill," asserting that they were "not indicting" women who had aborted on those grounds, and acknowledging that the matter had not yet been debated within either the women's movement or the disability movement, the authors nonetheless staked out a clear stance in opposition to the eugenic indication. "Wir Krüppelfrauen und der Paragraph 218," in Boll et al., *Geschlecht*, 80. It was this latter position that was to have enduring implications.

53. Incredibly enough—and while three hundred counterdemonstrators, including Hermes herself, along with many other individuals with disabilities and representatives of feminist groups, had shown up at Hadamar the day of the demonstration in order to offer a different interpretation of the lessons of that place at which, during the Third Reich, fifteen thousand individuals had been murdered by gas and by poison—an agitated antiabortion activist woman had screamed at Hermes (who was there in a wheelchair): "'Why don't you kill yourself, you are not worthy of living. Under Hitler something like this would not have happened!'" (Gisel Hermes, "Mensch achte . . . ," *Die Randschau* 1, no. 3 [August/September 1986]: 11–12).

54. "Sind die Abtreibungsgegner noch zu retten?," *Die Randschau* 1, no. 3 (August/September 1986): 14–15.

55. Swantje Köbsell and Monika Strahl, "Recht auf behindertes Leben: Humangenetik, die 'saubere Eugenik' auf Krankenschein," *Die Randschau* 1, no. 3 (August/September 1986): 9–10. They reported further that while nondisabled feminists and leftists gladly criticize "the 'evil' state" and what they too see as "a means of control over women's bodies in order to enforce the population-political aims of those who rule" (which they too vigorously believe is "an incursion into a woman's right to self-determination"), the minute these same individuals were asked to call for or abide by abolition or boycott of prenatal diagnostics, "the personal-emotional side of the matter comes into play." These women "want to know whether their child will be disabled or not. Unfortunately, it is hardly possible to address this contradiction."

56. For instance, when Swantje Köbsell was invited by Green Party feminists

to address a hearing on abortion rights in the Bundestag in November 1988, she expressly declared: "We are for the right of the woman to get an abortion and decide whether she wants a child or not. But we are against the aborting of a fetus solely on the grounds of its deficient quality, at a point when the woman had actually already decided in favor of carrying the pregnancy to term" ("'Unwertes Leben' darf abgetrieben werden—Bevölkerungspolitik in der BRD," in *Bevölkerungspolitik und Tötungsvorwurf: Dokumentation zweier Foren der GRÜNEN Frauen im Bundestag. . . . 21. Nov. 1988 . . . 5. Dez. 1988,* by DIE GRÜNEN im Bundestag, Arbeitskreis Frauenpolitik [Bonn, 1989], 30). Köbsell reported as well that she and other disabled women felt "attacked in our own right to life" (*in unserem Lebensrecht getroffen*) when nondisabled women either defended or sought prenatal testing. Meanwhile, Köbsell also noted—and in this she was of course correct—that only a small minority of disabling conditions was even visible in prenatal tests, whether amniocentesis or ultrasound. She phrased it somewhat differently, saying that at most "5 to 10 percent" of all disabilities even *existed* before birth, with "the remaining 90 to 95 percent occurring during birth or after" (30). For her, this served as yet another reason to boycott prenatal testing. (Inevitably, those statistics have changed with the development of both preimplantation genetic testing for in vitro fertilizations and the noninvasive prenatal screening available for all pregnant women since 2011, but the proportion of disabling conditions that are legible in the genetic material remains low. Among other things, there are no genetic markers yet for autism, and, in view of both the blurrily bounded umbrella category that it is and the most recent epigenetic findings suggesting that environmental conditions can have an impact on gene expression, not to mention any number of random things that can go awry in the course of a pregnancy, the proportion is likely to remain low. There will always be disability.)

57. In his 1983 book, *Krüppelschläge: Gegen die Gewalt der Menschlichkeit* (Crutch strikes: Against the violence of humanitarianism), Christoph was especially vehement about "the almost matter-of-course prenatal prevention of cripple misery," which, in his view, "also affects us living cripples. . . . I am afraid at the thought that there is a growing readiness to prevent something like me from the beginning and even to present oneself as 'humane' in doing so." Christoph further confronted self-understood progressive defenders of abortions for reasons of disability with the imaginary comparison of arguing that "blacks in [apartheid] South Africa cannot be expected to burden themselves with having children, because these would be at the mercy of oppression" (35).

58. Christoph, *Krüppelschläge*, 35. Again in 1989, Christoph expressed outrage that the feminist movement was demanding abortion rights but not coupling their demands with a stance in clear favor of disability rights and against "eugenic selection." Christoph quoted in "Krüppelschläge: Wie weit reicht das Selbstbestimmungsrecht der Frau?," *Konkret* 4 (1989): 41–48, here 42.

59. Peter Rödler, editorial in *Behindertenpädagogik* 29, no. 1 (January 1990): 2–6, here 5. This remark by Rödler is also quoted in Susanne Ehrlich, *Denkverbot als Lebensschutz? Pränatale Diagnostik, Fötale Schädigung und Schwangerschaftsabbruch* (Opladen: Westdeutscher Verlag, 1993), 195. Ehrlich provides an original and provocative argument about the just-then-growing hostility to women who sought prenatal diagnosis and, after having done so and learned of an anomaly, chose to terminate their pregnancies. Building on prior work by her advisor, the sociologist Gerhard Amendt, who had studied the ascent of the accusation that abortion was equivalent to murder, Ehrlich proposed that unconscious rage and fear at the prospect of a woman refusing to be an unconditionally loving and self-giving mother fueled much antiabortion activism and was especially exacerbated in the case of a disability pregnancy. See also Gerhard Amendt, *Die bestrafte Abtreibung: Argumente zum Tötungs-vorwurf* (Bremen: Ikaru, 1988); and Amendt with Michael Schwarz, *Das Leben unerwünschter Kinder* (Frankfurt am Main: Fischer, 1992).

60. Susanne von Paczensky et al. in the roundtable published as "Krüppel-schläge." Christoph noted that this was "the first time" such a discussion about the competing pulls of abortion rights and disability rights had taken place.

61. The one major feminist venue that ran an essay defending Singer and also defending abortion on grounds of fetal disability—the magazine *Emma*—would find its offices the object of a rampage when women in monkey-face masks destroyed 100,000 Deutschmarks' worth of office equipment and spray-painted the walls with "Emma engages in selection!," "Enough with the racism!," and "Euthanasia is violence." See "Islamismus: Der Überfall," *Emma*, July/August 1994. *Emma*'s countercharge of "Islamism" was an (obviously racially inflected) barb directed at the intruders to convey what *Emma* saw as their fundamentalism and intolerance. The original offending article was Filter, "Das Affentheater."

62. *Konkret* image reprinted in Udo Sierck, *Das Risiko nichtbehinderte Eltern zu bekommen: Kritik aus der Sicht eines Behinderten* (Munich: AG SPAK, 1989), 77. The New Left journal *Konkret* was the most offensive (although the Berlin-based daily *tageszeitung* was not far behind, and a well-known feminist cartoon-ist also contributed to the derogatory representations of disability). In addition

to the 1985 thalidomide item, in 1986, in the wake of the Chernobyl reactor disaster, the prominent and beloved *Konkret* cartoonist Ernst Kahl, in his regular "children's corner," drew an elaborate cartoon that commented on Ronald Reagan's "Star Wars" Strategic Defense Initiative, launched in 1983 (with a reference to Reagan's defense secretary, Caspar Weinberger), and simultaneously on the Chernobyl radiation fallout. Under the opening remark that "more and more children are coming into the world with malformations. Some piece is missing, or it's stuck in the wrong place. But everything has its good sides, as you'll soon see," the cartoon went on, in the guise of describing children's games, to show how "everything's not so bad." Thus, "Peter" sits in a glass jar, and his sister can show him off and carry him around. "Nelli" can win at blindman's bluff because she has a third eye, on her forehead, and can see above the blindfold. "Wotan" can go snorkeling without a snorkel because his nose sits on top of his head. "Tom" can play Indian because he has a feather growing out of his head; while "Bärbel" can save the entrance fee to the zoo, since when she looks in the mirror, her own face is that of a monkey's. And "Dragon"—clutching a little American flag in his baby hands—simply looks blank: "He has nothing in his head. He just smiles all the time. He's not understanding anything. Everyone is jealous of him!" Along related lines, the feminist cartoonist Franziska Becker published a cartoon in the feminist women's magazine *Emma* in 1986 in which an indecipherable blob in a playpen is being spoonfed while a visitor asks the father, "I don't know, does he look more like you, my boy, or more like Susanne?" These two images were published and criticized by Köbsell and Strahl, "Recht." In 1989 the *tageszeitung* published a cartoon in which talking monkeys in a jungle discussed why a particularly "ugly" monkeychild had been born ("a mutation . . . Since the start-up of the nuclear power plant, that's been happening . . . more frequently"). Reprinted in Sierck, *Das Risiko*, 72.

63. On some nondisabled Green feminists' extraordinary commitment to sensitivity and solidarity with disabled feminists, see Verena Krieger, "Selbstbestimmung der Frau—eine grundsätzliche Debatte," in DIE GRÜNEN, *Bevölkerungspolitik*, 9–13; and the brilliant overview of the state of debate over how best to defend abortion rights in Verena Krieger, "Die neue Abtreibungsdebatte in der Frauenbewegung," *Blätter für deutsche und international Politik*, 3, no. 365 (1989): 3–10. The quote is from Maria Mies, who—notably evincing a different kind of insensitivity—went on to argue that reproductive technologies represented a "'new eugenics on a global scale' that would make Hitler's racial politics seem like mere 'child's play.'" Quoted and discussed in Kimba

Allie Tichenor, *Religious Crisis and Civic Transformation: How Conflicts over Gender and Sexuality Changed the West German Catholic Church* (Lebanon, NH: UPNE, 2016), 202. The shared suspicion—once again, the embryo was being detached from the living woman and made an object for male doctors, said some otherwise prochoice feminists—and the desire to show sensitivity and solidarity especially toward feminist women with disabilities were major factors hampering the development of arguments that distinguished between abortion and murder. Suspicion of doctors made sense in a post-Nazi nation, in which indeed there were numerous continuities of personnel and attitude across the divide of 1945; feminist objections to the intrusion of science into natural processes and the prospect that embryos could be used for research purposes or meddled with in any way converged with broader environmentalist trends. Not every feminist was so ardently opposed to reproductive technologies that she would, as some did, break into the offices of genetic counseling centers and smash the equipment. (*Die Randschau* reported on one such incident.) But—as became clear at the latest in the protests against Singer—there were already in existence dozens of feminist organizations with such names as Rhein-Main Rats against Gen- and Reprod-Nonsense, Women against Genetic and Reproductive Technology (from Darmstadt, Frankfurt, and Mainz), along with Mixed[-Gender] Group against Genetic and Reproductive Technology, and the Cripple Women Group against Genetic and Reproductive Technologies and Eugenics West Berlin. See "Wider den tödlichen philosophischen Liberalismus," *die tageszeitung*, January 10, 1990. Apparently, the "Women against Genetic and Reproductive Technology" conference held in Bonn in 1985 was the first time that a public call was made for "abolition or boycotting of genetic counseling." (Green Party feminists were co-organizers of this conference; approximately two thousand women from Germany and abroad attended.) Tichenor, *Religious Crisis*, 201.

64. Köbsell, "'Unwertes Leben.'" For an excellent exposition and analysis of the multiple reasons for Green feminists' deep skepticism about reproductive technologies of all kinds, see Kristen Loveland, "Feminism against Neoliberalism: Theorising Biopolitics in Germany, 1978–1993," *Gender & History* 29, no. 1 (April 2017): 1–20.

65. "Wider den tödlichen philosophischen Liberalismus," also reprinted in *Die Randschau* 5, no. 1 (January–April 1990): 24. For Singer's defenders to complain that protests against him had become violent was, the undersigned declared, "an inversion of the actual violence relations." The life right of the disabled, they contended, should not be "discussable." They also pinpointed,

as a sign of just how disability-hostile the society was, the way that "genetic counseling and the prenatal selection of disabled fetuses is treated by doctors and jurists as 'responsible pregnancy precaution.'" And they averred that "whoever does not participate in this 'voluntary eugenics,' is increasingly discriminated against."

66. It may be worth noting here that in 1983 Catholic Church leaders had been distressed to learn that an opinion survey showed West Germans considered the killing of baby seals a more serious moral offense than abortion; taking their cue from the rising popularity of the Green Party, Catholic conservatives learned to use environmentalist arguments in their antiabortion campaigns. By 1989 they had additionally picked up on the disability rights movement's objections to terminations based on prenatal testing. In a document cowritten with Protestant religious leaders, "Gott ist ein Freund des Lebens" (God is a friend of life), endorsed in November 1989 by the head of the German Council of Bishops, they argued, in an extended discussion of many dimensions of disability rights, that "finally, we cannot overlook the mentality that could develop concerning the life of disabled persons and their acceptance by society as a result of the coupling of prenatal diagnosis and abortion. . . . Society could reach a point where it no longer accepts disabled children. They need not have been born. For the self-image of the disabled, the consequences would be incalculable, given such an assessment by the world around them." Never before, as the historian Kimba Allie Tichenor has written, had the German bishops "framed their objection to abortion in terms of a violation of the rights of the disabled." Discussion and quote translation in Tichenor, *Religious Crisis*, 205; the document is available at https://www.ekd.de/gottistfreund_1989_mitarbeiter.html. The self-evidence with which fetal abnormality had once been seen as an understandable, morally permissible grounds for termination—not, certainly, by the Catholic Church leadership but by the general populace—is evident in statistics from the early 1970s. In a 1971 survey, 83 percent of Germans and *80 percent of Catholics* who were questioned stated their conviction that this was so (this in contrast to 54 percent of the total population and 51 percent of Catholics who thought women should have an unconditional right to an abortion for any reason). Tichenor, *Religious Crisis*, appendices.

67. The 1980s saw an efflorescence of conservative politicians' efforts to recriminalize abortion more completely. Different strategies were being employed, among them the effort to declare that only abortions undergone on grounds of danger to maternal health should be covered by insurance companies; more than fifty Christian Democratic and Christian Social politicians had put

forward a formal proposal in the Bundestag that abortions on grounds of the "social indication"—the most widely used one—should be paid for by the woman herself; at the same time, they launched an initiative, "Mutter und Kind" (Mother and child), that aimed to give some financial assistance to women willing to carry their pregnancies to term. This initiative—as small as the sums were—had substantial symbolic importance not least because politicians made the claim that abortions needed on social indication grounds were harder to justify. At the same time, to add to the ironies, social services were more generally suffering cutbacks. See "Die Reform des Paragraph 218 in Gefahr," in Boll et al., *Geschlecht*, esp. 76–78. By 1989 feminist politicians in the Green Party were sufficiently alarmed by the rollback apparently under way that they organized a series of hearings in the Bundestag in order to rearticulate, for legislators' sake, the reasons that Paragraph 218, which regulated abortion, should be abolished entirely and to rebut what was called, in the terminology of the time, the "murder charge" (*Tötungsvorwurf*)—the accusation that abortion and murder were indistinguishable. The Green Party, once conservative environmentalists had split away, had become *the* voice, in the West German party-political arena, in defense of abortion rights—and took heat from the other parties for that reason alone.

68.   Bavaria had long been a conservative outlier among the western states, and a court case brought in 1988–89 against a doctor who had provided abortions to hundreds of desperate women in the Bavarian town of Memmingen had riveted the nation's attention as dozens of his female patients, called as witnesses, were subjected by the court to a glaringly public, devastating intrusion into the most intimate aspects of their private lives, the quality of their sexual and romantic relationships, and the exigencies of their financial and familial situations. This intrusion occurred supposedly because the state's attorney and the court were concerned that perhaps in some cases there had not been a sufficiently emergency situation that the granting of an abortion on grounds of the social indication had been warranted—and not least because the most aggressive interrogator among the judges had been revealed to have taken his own girlfriend to get an abortion in the neighboring and more liberal state of Hessen. The sensational scandal of a double standard aside—and ultimately this judge had to recuse himself—the Memmingen trial had made starkly, frighteningly clear to women everywhere just how eroded abortion rights had become. Quick entries into the emotional intensities of the Memmingen case are provided by Uta König, "[Paragraph] 218," *Stern*, January 26[?], 1989, 12–13; Gerhard Mauz, "'Hätten Sie es nicht in Pflege geben können?,'" *Spiegel*, March 13,

1989; "Gnadenlos," *Die Zeit*, March 17, 1989; and Monika Frommel, "Der Kreuzzug von Memmingen," *Neue Kriminalpolitik* (1990). In 1990, already before unification, the Christian Social Union–led Bavarian state government had asked the Constitutional Court for a judicial review of the West German abortion law. After unification and after the Bundestag passed a new law in 1992, against the objections of 248 (in some accounts 249) Christian Democratic and Christian Social Union parliamentarians, these individuals, together with the Bavarian state government, petitioned the Constitutional Court for a temporary injunction to be issued; the Constitutional Court provided that, and then promulgated its own decision in 1993, relying also on the Bavarian state's arguments. These among other things revealed a patronizing didactic intention vis-à-vis the East Germans. If the new version of Paragraph 218 were to be found to be unproblematic from a constitutional point of view, the Bavarian petition asserted, "then the concept of the trimester solution prevailing since 1972 in the accession region [*Beitrittsgebiet*, i.e., the territory of the former GDR] would ultimately be affirmed. The legislature would thereby be forfeiting the opportunity to, with its means, foster a lawful consciousness for the worth of unborn life and its constitutional protection in the populace of the new federal states." The court decision contains substantial passages from the Bavarian statement: Antrag auf verfassungsrechtliche Prüfung von Vorschriften über das Beratungs- und Indikationsfeststellungsverfahren sowie über Krankenversicherungsleistungen bei Schwangerschaftsabbrüchen aufgrund der allgemeinen Notlagenindikation (2 BvF 2/90). Constitutional Court decision and accompanying commentary, BVerfGE 88, 203—Schwangerschaftsabbruch II (see esp. pp. 219, 238, and quote at 240), can be found at http://groups.csail.mit.edu/mac/users/rauch/nvp/roe/bvo88203_nonav.html. Further condescending remarks about the East German populace had been put forward during lawyers' presentations to the Constitutional Court. The jurist Peter Lerche, for instance, noted—as summarized by the *Frankfurter Allgemeine Zeitung*—that "deep wounds" had been created in "the sense of right and wrong" prevailing in "the new federal states" (i.e. the states of the former GDR) due to the "twenty years there of unconditional acceptance of the trimester regulation." Quoted in "Das Bundesverfassungsgericht erlässt einstweilige Anordnung gegen die Fristenlösung: Zweiter Senat gibt den Anträgen Bayerns und von Abgeordneten der Union statt," *Frankfurter Allgemeine Zeitung*, August 5, 1992.

69. For a virtuoso critique of the Constitutional Court's bald-faced and highly consequential shift in focus, in 1992–93, from the living human beings (*Menschen*) guaranteed protection by the German *Grundgesetz* to the defense

of a new—vague and abstract—category they referred to as "life" (*Leben*), see Barbara Duden, "Nachwort zum Karlsruher Urteil: 'Das Leben' als Entkörperung," in *Der Frauenleib als öffentlicher Ort*, 2nd ed. (Munich: dtv, 1994), 147–63.

70. In general, the hope was that working with women (through counseling), rather than against them (through punishment), would encourage more women to continue their pregnancies. The Bavarian petition of 1990 had already, in its panoply of arguments for restricting abortions, floated the worry that "improvements in prenatal diagnostics" could eventually make "abortion on purely eugenic grounds become possible," potentially already "within the first twelve weeks"—such that it would be impossible to know whether "the continuation of the pregnancy" could have been "demanded of the woman." See BVerfGE 88, 203—Schwangerschaftsabbruch II, p. 240. But the proposed law of 1992 had still included the idea that an abortion should be permissible in cases in which "in accordance with medical finding there are pressing reasons for assuming that the child would suffer irreparable damage to its health, due to a genetic inheritance or toxic effects during the pregnancy, that weigh so heavy that the continuation of the pregnancy cannot be demanded of the woman" (p. 228; cf. p. 221). Although including mention of these two positions, more generally, the embryopathic indication, pro or con, was not a matter that the court in 1993 had particularly belabored. The court's own stance consisted of the remark that

> the untenability of continuing a pregnancy, however, cannot arise from circumstances that remain within the framework of the normal situation of a pregnancy. On the contrary, there must be burdens that demand such a degree of self-sacrifice that this cannot be expected of the woman. In view of the duty to carry the child to term, it follows that, beside the traditional medical indication, the criminological and, insofar as the boundaries are drawn sufficiently precisely, the embryopathic indication can exist as exceptions that are in accordance with the constitution; for other situations of emergency this is only the case if the severity of the social or psychological-personal conflict presupposed here is clearly discernible. (p. 258; cf. pp. 270, 300)

Bottom line: at this point the court was still most worried about reining in the far larger number of "social" cases. Nonetheless, signs of a wider shift in conceptualizations of the embryopathic indication were already evident among jurists in 1993. In the wake of the court's decision, conservative legal experts interested in an at once restrictive and pragmatic resolution (pragmatic both because, as the court had also been aware, "abortion has been and remains a mass phenomenon" and because one would not want to "intensify the resistances, especially in the new federal states"—the latter again a discreetly phrased but pointed gesture to potential unrest in the former East) were attuned to the idea that it could be important, if the embryopathic indication were to remain, that its inclusion be articulated in such a way that "not the disability of the child as such" would legitimate abortion but only whether, in a given individual case, that disability would truly be too difficult for the woman in question to manage. Anticipated disability alone should no longer be misunderstandable as somehow an acceptable reason to terminate. Rolf Keller, "Das Urteil des Bundesverfassungsgerichts zum Schwangerschaftsabbruch vom 28. Mai 1993 aus strafrechtlicher Sicht: Zur Rolle des Strafrechts im Beratungskonzept," in *Paragraph 218, Urteil und Urteilsbildung,* ed. Johannes Reiter and Rolf Keller (Freiburg: Herder, 1993), 195–216, here 204, 208–9. As of 1994, the SPD's proposed version of the law, without getting into any detail or explaining its reasoning, had suggested the subsumption of the embryopathic indication in the medical indication. See "Gesetzentwurf der Fraktion der SPD: Entwurf eines Gesetzes zur Anpassung des Schwangeren- und Familienhilfegesetzes an die Vorgaben des Urteils des Bundesverfassungsgerichts vom 28. Mai 1993," Deutscher Bundestag, 12. Wahlperiode, Drucksache 12/6669 (January 25, 1994): 3–4, http://dipbt.bundestag.de/doc/btd/12/066/1206669.pdf.

71. "Beschlussempfehlung und Bericht des Ausschusses für Familie, Senioren, Frauen und Jugend (13. Ausschuss), zu dem 1. Gesetzentwurf der Fraktion der CDU/CSU—Drucksache 13/285—Entwurf eines Schwangeren- und Familienhilfeänderungsgesetzes (SFHÄndG), 2. Gesetzentwurf der Fraktion der SPD—Drucksache 13/27 . . . ," Deutscher Bundestag, 13. Wahlperiode, Drucksache 13/1850 (June 28, 1995): 16, http://dip21.bundestag.de/dip21/btd/13/018/1301850.pdf. Only the Party of Democratic Socialism (the legal successor to the former East German Socialist Unity Party) protested, eloquently, against the abrogation of women's rights to seek terminations on any grounds, accusing the Constitutional Court of "intolerable paternalism" and declaring that the court itself, in its decision of May 1993, had "restricted the fundamental rights of women to the inviolability of their dignity, the free development of their personality, and their freedom of conscience for the full duration of the pregnancy" (18–19).

72. On February 10, 1995, after the proposed law had been read the first time, the committee was charged to deliberate on it—Hüppe was part of this committee. The others were Maria Eichhorn (CSU and a member of Donum Vitae), Heinz Lanfermann (FDP), Inge Wettig-Danielmeier (SPD), Rita Griesshaber (B90/Greens), Christina Schenk (PDS, was lesbian, is now

Christian Schenk, a trans man). One earlier version of a CDU/CSU proposed law—before the Constitutional Court had gotten involved—had already suggested the possibility of subsuming cases of rape (the criminological or "ethical" indication), of fetal anomaly (termed "child's indication"), and of a "crisis situation" (*Notlage*) under the rubric of a "psycho-social" indication which would be handled separately from the maternal health indication. See "Empfehlung und Bericht des Sonderausschusses 'Schutz des ungeborenen Lebens,'" Deutscher Bundestag, 12. Wahlperiode (June 22, 1992): 93–94, http://dipbt.bundestag.de/doc/btd/12/028/1202875.pdf.

73. In addition, in a positive development, the counseling component of the law would come to include, in instances of fetal disability, information about "the options of assistance for disabled people and their families that are available before and after birth" and the right to access experts on early intervention with disabled children. Interestingly, moreover, the commentary went on to explain who had been most inspirational in the decision causing the embryopathic indication to be "dispensed with": "Above all the statements of disability organizations had namely shown that such a regulation had led to the misunderstanding that the legitimation [of a termination in these circumstances] follows from a reduced valuing of the life rights of a disabled child. . . . Herewith is made clear that a disability can never lead to a diminution in protection of life." See "Beschlussempfehlung," 5–6, 18, 20, 25–26.

74. See "Beschlussempfehlung," 19; and "Antwort der Bundesregierung auf die kleine Anfrage der Abgeordneten Hubert Hüppe et al.," Drucksache 13/5364, July 29, 1996, http://dipbt.bundestag.de/doc/btd/13/053/1305364.pdf. See also the careful statement of the Federal Chamber of Physicians advising doctors how to work in accordance with the law of October 1995: "Erklärung zum Schwangerschaftsabbruch nach Pränataldiagnostik," *Deutsches Ärzteblatt*, November 20, 1998, A-3013–A-3016.

75. As Tichenor succinctly observes, "Green Party and Social Democratic women did not want to be seen as opponents of disability rights" (*Religious Crisis*, 206–7). As of 2009, officially 3,200 of a total number of 110,694 abortions were undertaken under the rubric of the medical indication; by 2015 it was 3,879 out of 99,237. Statistisches Bundesamt, *Schwangerschaftsabbrüche* (2016), 29–30, https://www.destatis.de/DE/Publikationen/Thematisch/Gesundheit/Schwangerschaftsabbrueche/Schwangerschaftsabbrueche.html. Commentators across the ideological spectrum assume that a substantial majority of these were undertaken on the basis of a diagnosed embryopathic condition, but many also have some skepticism about the accuracy of the statistics. Notably,

in Austria in 2015, it was the Green Party—invoking the UN Convention on the Rights of Persons with Disabilities (even though it does not mention abortion at all)—that urged the Austrian parliament to consider whether a shortened time frame for aborting on grounds of fetal anomaly might be appropriate. The conservative Austrian People's Party concurred that it had "long been in favor of removing the eugenic indication," since the existence of that indication in Austrian law represented an "unequal treatment of disabled people." See "Spätabtreibungen: Debatte über Fristverkürzung," *Der Standard*, February 9, 2015, http://derstandard.at/2000011464441/Spaetabtreibungen-Debatte-ueber-Fristverkuerzung.

76. Hüppe has among other things argued that the (partial) legalization of preimplantation diagnostics in Germany is in contradiction to the UN Convention on the Rights of Persons with Disabilities. See "Behindertenbeauftragter kritisiert PID-Regelung," Stoppt PID website, http://www.stoppt-pid.de/bei traege/behindertenbeauftragter_kritisiert_pid-regelung.

77. Kay-Alexander Scholz, "NS-Euthanasie: Der Probelauf zum Holocaust," *DW*, February 1, 2013, http://www.dw.com/de/ns-euthanasie-der-probe lauf-zum-holocaust/a-16569574.

Chapter 3. Time Well Wasted

1. In the first effort at global assessment, and while noting that "disability, a complex multidimensional experience . . . poses several challenges for measurement," the World Health Organization and the World Bank nonetheless declared summarily in 2011 that "more than one billion people [of the seven billion people] in the world live with some form of disability, of whom nearly 200 million experience considerable difficulties in functioning. In the years ahead, disability will be an even greater concern because its prevalence is on the rise. This is due to ageing populations and the higher risk of disability in older people as well as the global increase in chronic health conditions such as diabetes, cardiovascular disease, cancer and mental health disorders" (World Health Organization, *World Report on Disability* [Geneva, 2011], xi, 21). The European Union has a total population of 508 million inhabitants (this is the world's third largest population after China and India), and the European Union Agency for Fundamental Rights announced in 2010, citing the European Commission Disability Strategy 2010–2020, that about 80 million of these people have a disability (and more than 1 million of them live in an institution). See FRA, "The Fundamental Rights of Persons with Intellectual Disabilities

and Persons with Mental Health Problems" (2010), http://fra.europa.eu/sites /default/files/fra_uploads/1292-Factsheet-disability-nov2010.pdf. Taking into account all kinds of disabilities, it is estimated that one-sixth of Europe's working-age population (approximately 42 million) is disabled. See European Commission, "Persons with Disabilities," http://ec.europa.eu/social/main .jsp?catId=1137&langId=en; and Eurostat, "Disability Statistics—Prevalence and Demographics" (2015), http://ec.europa.eu/eurostat/statistics-explained /index.php/Disability_statistics_-_prevalence_and_demographics. On the extraordinary importance of—but also the inevitable problems with—attempting to quantify disability, see Faye Ginsburg and Rayna Rapp, "Making Disability Count: Demography, Futurity, and the Making of Disability Publics," *Somatosphere*, May 11, 2015, http://somatosphere.net/2015/05/making-disability-count -demography-futurity-and-the-making-of-disability-publics.html. Ginsburg and Rapp quote a US Institute of Medicine document from 2007 that aptly observes: "If one considers people who now have disabilities, people who are likely to develop disabilities in the future, and people who are or who will be affected by the disabilities of those close to them, then disability affects today or will affect tomorrow the lives of most Americans. The future of disability in America is not a minority issue."

2. For example, see the "E allora?" (So?) campaign in Italy in 2003, with the slogan "If you could feel like me, you would understand the beauty of life." The posters are available at http://www.pubblicitaprogresso.org/schede_ mediateca/e-allora/ and the video at https://player.vimeo.com/video/59153307. Along related lines is Charita Opava's "No a co?" (So what?) campaign for meaningful employment for individuals with various physical or cognitive disabilities in the Czech Republic in 2012. Some of the posters are available at http://www.charitaopava.cz/?page=aktuality&id=895 and http://www.nfozp .cz/nase-projekty/ukoncene-projekty/kampan-no-a-co/. See also the "We will not let ourselves be disabled" campaign by Pro Infirmis in Switzerland, which ran from 2001 to 2008, with posters available at http://www.proinfirmis.ch /medien/download/kampagnen-2001-2008.html. And for an example of bodily difference celebrating disability awareness activism, see the mannequin video created by Pro Infirmis in 2013, https://www.youtube.com/watch?v=E8um FV69fNg. The slogan is "Because who is perfect?"

3. For instance, for France see "Vaincre l'autisme," http://www.vaincrelau tisme.org/; at the website for the UK's National Autistic Society you can play a quiz guessing the identities of famous people rumored to be autistic or to have Asperger's syndrome (they include Thomas Jefferson, Albert Einstein, Dan

Aykroyd, Courtney Love, and Wolfgang Amadeus Mozart). See http://www
.autism.org.uk/get-involved/world-autism-awareness-week/resources.aspx.
For the Down Syndrome Association of Greece, see http://www.down.gr/; for
Croatia, see Roberta Mesic, "Buba Bar—Ekstra kromosom za ekstra kavu" (a
coffee shop employing individuals with Down syndrome under the headline
"Extra chromosome for extra coffee"), https://www.indiegogo.com/projects
/buba-bar-ekstra-kromosom-za-ekstra-kavu#/. Also in Croatia, government
leaders participated in the campaign to wear mismatching socks on World Down
Syndrome Day. See "Šarene čarape na nogama hrvatskih političara i politi-
čarki," http://dnevnik.hr/vijesti/hrvatska/i-andrej-plenkovic-obiljezio-svjetski
-dan-osoba-s-downovim-sindromom-470524.html. For a rousing British defense of the meaningful and rich lives of individuals with Down syndrome—
including discussion of the achievements and arguments of US activist Karen
Gaffney, who has Down syndrome, as well as of the views of British actress and
screenwriter Sally Phillips, whose son Olly has the condition—along with
images from Icelandic photographer Sigga Ella's project representing Icelandic
individuals with Down syndrome, see Alison Gee, "A World without Down's
Syndrome?," *BBC News*, September 29, 2016, http://www.bbc.com/news/maga
zine-37500189. The United States, however, still tops the charts with slogans
such as "I may not make eye contact but at least I don't stare" and "God created
autism to offset the excessive number of boring people on earth," as well as
"Down syndrome happens randomly, like flipping a coin or winning the lot-
tery," "Keep calm, it's only an extra chromosome," "Down syndrome is not a
tragedy, ignorance is a tragedy," "Don't you dare underestimate me . . . I
wouldn't do that to you," and "Prenatal testing cannot predict this kind of
love." See https://www.pinterest.com/americanautism/inspired/ (accessed July
14, 2017); https://www.pinterest.com/pin/428897564484273899 (accessed July
14, 2017); and http://www.jillstanek.com/2013/01/billboard-prenatal-testing
-cannot-predict-this-kind-of-love/.

4. For example, for Germany, see Aktion Mensch's "Das erste Mal,"
https://www.youtube.com/watch?v=gZFHK3OwzFM. This may be the right
place to note that I deliberately use the oft-recommended "person-first" language in which one speaks of "individuals [or persons] with disabilities" inter-
changeably with such terms as "the disabled" or "disabled individuals." For
explanations of the value of the latter terms not least because they convey the
double meaning within the word "disabled" (signaling not solely the impairment
in the individual but also, and importantly, the disablement caused or exacerbated by social conditions), see Don Kulick and Jens Rydström, *Loneliness and*

Its Opposite: Sex, Disability, and the Ethics of Engagement (Durham, NC: Duke University Press, 2015), 34; and Sigrid Graumann, *Assistierte Freiheit: Von einer Behindertenpolitik der Wohltätigkeit zu einer Politik der Menschenrechte* (Frankfurt: Campus, 2011), 12.

5. On Best Buddies in the UK, see https://www.bestbuddiesuk.org/; for the celebrity ambassadors on airport walls, see https://bestbuddies.org/celebrity-ambassadors/; and for programming in the United States, see https://bestbuddies.org/what-we-do/friendship/. On the Wounded Warrior Project, see https://www.woundedwarriorproject.org/.

6. Just a few examples: on the 2004 British film *Every Time You Look at Me*, a love story starring Mat Fraser, an actor affected by thalidomide, and Lisa Hammond, an actress with restricted growth, see http://www.bbc.co.uk/press office/pressreleases/stories/2004/03_march/19/every_time.shtml. The German theater troupe Ramba Zamba relies on actors and actresses endowed with Down syndrome. See http://www.theater-rambazamba.org/. Poland too has a theater troupe, Teatr Razem, whose actors include "persons with mental disabilities." See http://psoni.gda.pl/teatr-razem/. Poland in 2015 sent Monika Kuszyńska, who uses a wheelchair, to the Eurovision contest. See https://www.youtube.com/watch?v=TReYIZYloSg.

7. For a real-life love story between a nondisabled woman (Karin Knoll) and a man who had contracted polio at age two and lived for many years in an iron lung (Ferdinand Schiessl), see "Ferdinand Schiessl: Ich bin der Frosch, küss mich!," *Rolling Planet*, November 9, 2013, http://rollingplanet.net/ferdinand-schiessl-ich-bin-der-frosch-kuess-mich/; for another real-life love story, this one between a woman with cerebral palsy (Bernadette Gradl) and a man with restricted growth (Ulrich Ochs), see Nina Job, "Handicap-Familie: 'Was besseres konnte uns nicht passieren,'" *Abendzeitung* (Munich), January 18, 2016, http://www.abendzeitung-muenchen.de/inhalt.mutter-und-vater-mit-behinderung-handicap-familie-was-besseres-konnte-uns-nicht-passieren.ff70e512-e3fa-4998-a0de-a50039dab9d3.html. For internationally publicized erotic photography of individuals with physical disabilities and differences staged in Italy by the photographer Olivier Fermariello, see Ellyn Kail, "Intimate Photos Take Us into the Bedrooms of People with Disabilities," *Feature Shoot*, August 12, 2014, http://www.featureshoot.com/2014/08/sensual-photos-take-us-into-the-bedrooms-of-people-with-disabilities-nsfw/. A terrific overview and interpretation of emergent trends in awareness and publicity is provided by Ginsburg and Rapp, "Making Disability Count."

8. On the evolution of debates and the key issues, see Jean-Noël Missa and

Charles Susanne, eds., *De l'eugénisme d'État à l'eugénisme privé* (Paris: De-Boeck, 1999); Michael J. Sandel, "The Case against Perfection: What's Wrong with Designer Children, Bionic Athletes, and Genetic Engineering," *Atlantic Monthly*, April 2004, 50–62; Hilary Rose, "Eugenics and Genetics: The Conjoint Twins?," *New Formations* 60 (Spring 2007): 13–26; Nikolas Rose, *The Politics of Life Itself: Biomedicine, Power, and Subjectivity in the Twenty-First Century* (Princeton, NJ: Princeton University Press, 2007); and Clare Hanson, afterword in *Eugenics, Literature, and Culture in Post-war Britain* (New York: Routledge, 2013), 149–58. For a different perspective on the trends of the neoliberal present and important reflections on how disability as a phenomenon inevitably puts pressure on questions of states' relationships to wealth distribution, see Robert McRuer, "Disabling Sex: Notes for a Crip Theory of Sexuality," *GLQ* 17, no. 1 (2011): 107–17; and David T. Mitchell and Sharon L. Snyder, "From Liberal to Neoliberal Futures of Disability: Rights-Based Inclusionism, Ablenationalism, and the Able-Disabled," in *The Biopolitics of Disability: Neoliberalism, Ablenationalism, and Peripheral Embodiment* (Ann Arbor: University of Michigan Press, 2015).

9. An outstanding analysis and overview of the German debates and what she notes is the "polyvalence" of terms like "dignity" and "eugenics" is provided by Kristen Loveland, "Re-producing the Human: Dignity, Eugenics, and Governing Reproductive Technology in Germany" (PhD diss., Harvard University, 2017). See also the classic statement by Jürgen Habermas, *The Future of Human Nature* (Cambridge: Polity Press, 2003); and the essays collected in Martin Stingelin, ed., *Biopolitik und Rassismus* (Frankfurt am Main: Suhrkamp, 2003). German public health organizations like Pro Familia (https://www.pro familia.de/) and the Arbeitskreis Frauengesundheit (http://www.akf-info.de /portal/) have taken on board the worry that the availability of new technologies puts pressure on individuals and constricts rather than expands their freedom of choice. The language used can at times be harsh, with explicit references to the Nazi past, as commentators make their case for contending that there has been no unlearning of eugenics. For instance, the legal expert Ute Sacksofsky, in a memorandum titled "The Constitutional Status of the Embryo in Vitro," developed for a German parliamentary inquiry, "Law and Ethics of Modern Medicine," in the fall of 2001, asserted:

> Assuming there were a decision for born people comparable to the one for preimplantation diagnostics, the parallel to the worst crimes of the National Socialist regime would be obvious. For the core of the matter is the decision to deny a person the right to life based on certain characteristics. The image of the

selection ramp at Auschwitz forces itself on us, just as does the "euthanasia program" carried out against people with disabilities. The fundamental decision about whether the life of a particular person is "worthy of life" or "unworthy of life" is left to another person who moreover turns their verdict into action, that is into the killing of "unworthy" life. One cannot imagine a greater "essential calling-into-question of the subject quality" of a human being. Ute Sacksofsky, "Der verfassungsrechtliche Status des Embryo in vitro— Gutachten für die Enquete-Kommission des Deutschen Bundestages 'Recht und Ethik der modernen Medizin,'" September 2001, http://publikationen .ub.uni-frankfurt.de/frontdoor/index/index/docId/3330. One of the peculiar features of the German-language discussions is the prevalence of the assertion that somehow the matter is being silenced. See, for instance, comments by the German early modern historian Maren Lorenz in 2016: "The ethical implications of prenatal diagnostics stand like an elephant in the room, but hardly anyone talks about them" ("Kommentar: 'Wir sind mittendrin in der Diskussion um Menschenzucht,'" March 29, 2016, http://news.rub.de/wis senschaft/2016-03-29-kommentar-wir-sind-mittendrin-der-diskussion-um -menschenzucht).

    10. Hannah Arendt, *The Origins of Totalitarianism* (San Diego: Harcourt Brace, 1967), 296.

    11. See Mark Mazower, "The Strange Triumph of Human Rights, 1933–1950," *Historical Journal* 47, no. 2 (2004): 379–98; Samuel Moyn, "The Intersection with Holocaust Memory," in *Human Rights and the Uses of History* (London: Verso, 2014); Samuel Moyn, *Christian Human Rights* (Philadelphia: University of Pennsylvania Press, 2015); Stefan-Ludwig Hoffmann, "Human Rights and History," *Past and Present* 232 (2016): 279–310; and Stanley S. Herr, Lawrence O. Gostin, and Harold Hongju Koh, eds., *The Human Rights of Persons with Intellectual Disabilities: Different but Equal* (New York: Oxford University Press, 2003).

    12. For the "Conclusions" of the 1967 symposium, see International League of Societies for the Mentally Handicapped / Ligue Internationale des Associations d'Aide aux Handicapés Mentaux, "Legislative Aspects of Mental Retardation," Stora Skondal, Stockholm, June 11–17, 1967, https://mn.gov /mnddc/parallels2/pdf/60s/67/67-ILS-ILS.pdf.

    13. See United Nations, A Compendium of Declarations on the Rights of Disabled Persons, http://pf7d7vi404s1dxh27mla5569.wpengine.netdna-cdn .com/files/library/united-nations-un.pdf.

    14. International League of Societies for the Mentally Handicapped / Ligue

Internationale des Associations d'Aide aux Handicapés Mentaux, "Legislative Aspects of Mental Retardation," item II.2.b.1.

15. Eve Sedgwick, "Introduction: Axiomatic," in *Epistemology of the Closet* (Berkeley: University of California Press, 1990), 1, 40. Weeks after I had thought to invoke Sedgwick's contrast between minoritizing and universalizing as a useful tool for disability as well, I discovered that Rosemarie Garland-Thomson had cited this Sedgwick idea already in 2002. That feels confirmatory of my independent instincts, although I would also note that Garland-Thomson is thinking primarily about bodily disability, which allows her to emphasize the social constructionist version of disability to the exclusion of what might also be more frankly acknowledged as the—often traumatic— "real" of cognitive or psychiatric disability, which are not solely constructions. For example, she says: "The informing premise of feminist disability theory is that disability, like femaleness, is not a natural state of corporeal inferiority, inadequacy, excess, or a stroke of misfortune. Rather, disability is a culturally fabricated narrative of the body, similar to what we understand as the fictions of race and gender" (Rosemarie Garland-Thomson, "Integrating Disability, Transforming Feminist Theory," *NWSA Journal* 14, no. 3 [Fall 2002]: 5).

16. See in this context Dimitris Anastasiou and James M. Kauffman, "A Social Constructionist Approach to Disability: Implications for Special Education," *Exceptional Children* 77, no. 3 (Spring 2011): 367–84; and Mitchell and Snyder, *The Biopolitics of Disability*.

17. For articulate criticisms of the "we're all disabled" line of argument, see Ernst Klee, *Behindert: Über die Enteignung von Körper und Bewusstsein* (Frankfurt am Main: Fischer, 1980), http://bidok.uibk.ac.at/library/klee-behindert .html; and Julia Kristeva, "Liberty, Equality, Fraternity, and . . . Vulnerability," *WSQ* 38, no. 1–2 (Spring 2010).

18. On the "belated human right" to desired sexuality, as well as the right to freedom from sexual harm, see the important reflections of the Austrian legal expert Helmut Graupner, "Das späte Menschenrecht (Teil 1)—Sexualität im europäischen und österreichischen Recht," *Sexuologie* 11, no. 3–4 (2004): 119–39.

19. See Volkmar Sigusch, "Lean Sexuality: On Cultural Transformations of Sexuality and Gender in Recent Decades," *Sexuality & Culture* 5 (2001): 23–56; and Volkmar Sigusch, *Sexualitäten: Eine kritische Theorie in 99 Fragmenten* (Frankfurt: Campus, 2016). On transsexual bodies, see Susan Stryker, "My Words to Victor Frankenstein above the Village of Chamonix: Performing Transgender Rage," *GLQ* 1, no. 3 (1994): 237–54; and Jacqueline Rose, "Who

Do You Think You Are?," *London Review of Books*, May 5, 2016. On cyberspace, actual bodies, queerness, disability, and "intercorporeal generosity" (a term they borrow from Ros Diprose), see Margrit Shildrick and Janet Price, "Deleuzian Connections and Queer Corporealities: Shrinking Global Disability," *Rhizomes* 11/12 (Fall 2005 / Spring 2006), http://www.rhizomes.net/issue11 /shildrickprice/.

20. For example, see the special issue of *GLQ* edited by Robert McRuer and Abby L. Wilkerson: "Desiring Disability: Queer Theory Meets Disability Studies," *GLQ* 9, nos. 1–2 (2003). For a compelling argument that the current right-wing battles against sexual freedoms of all kinds are "not business as usual (i.e. another wave of backlash)" but rather better understood as "a coordinated transnational effort to undermine liberal values by democratic means," see Agnieszka Graff and Elżbieta Korolczuk, "Facing an Illiberal Future: Conceptualising the Polish Gender Crusade in a Transnational Context," presentation at the Twenty-Second International Conference of Europeanists, "Contradictions: Envisioning European Futures," Paris, July 8–10, 2015, http://www.academia .edu/16513239/Facing_an_Illiberal_Future._Conceptualising_the_Polish_gen der_crusade_in_a_transnational_context. For more on the "geopoliticization" of sexuality, see Dagmar Herzog, "What Incredible Yearnings Human Beings Have," *Contemporary European History* 22, no. 2 (May 2013): 303–17. For comparatively early and especially thoughtful and sensitive reflections on sexual rights as human rights for individuals with physical and with cognitive disabilities, see the brochures produced by the German organization Pro Familia: *Sexualität und körperliche Behinderung* (Frankfurt am Main, 1997) and *Sexualität und geistige Behinderung* (Frankfurt am Main, 1998). The latter one opens with the declarations "Sexuality belongs to a human being's personhood—human beings with disabilities are no exception here." Indeed: "Sexuality is not disabled" (4–5).

21. Kulick and Rydström, *Loneliness*, 6. See also the important book on sexuality for individuals with disability in the US context: Michael Gill, *Already Doing It: Intellectual Disability and Sexual Agency* (Minneapolis: University of Minnesota Press, 2015).

22. Kulick and Rydstrom, *Loneliness*, 21–22, 28, 35–36.

23. Kulick and Rydström, *Loneliness*, 16, 20–21, emphasis added.

24. Deulofeu finds it appalling that parents of disabled adults treat them like "little angels" and "keep them away from sex, as though they were asexual people, without sentiments or desires." Indeed, she has gone so far as to argue that she thinks "we should haul into court all those parents who hinder their

disabled children from enjoying sexual lives of their own." See Víctor-M. Amela, "'Los discapacitados tienen derecho al goce sexual,'" *La Vanguardia*, October 7, 2014, http://www.lavanguardia.com/lacontra/20140710/54410949970/los-discapacitados-tienen-derecho-al-goce-sexual.html. On her work, see also Elena Parreño, "El sexo de los ángeles," *El Periódico*, May 2, 2011, http://www.dincat.cat/reportatge-peri%C3%B3dico_78153.pdf.

25. Then the voice-over declares: "Sex is a serious matter. Sex is laughter, floating, melting." And another woman, in a wheelchair, expresses: "Sex is something natural, something that forms us and shapes us and is related to all aspects of our lives, sex in general, we are sexuate beings." The film is explicit, pornographic—but in a way that is appropriate to its ethically engaged, prosex message—as it covers a broad diversity of situations, including a man with cerebral palsy engaging an S/M prostitute for the first time (both are nervous), the use of sex toys, intercourse in different positions, gentle caressing, vigorous sensual massage, and kissing between people who have just met (sometimes one of them is a therapeutically trained "sexual assistant" or surrogate) and also between people who are long-term romantic life partners. All of this is replete with people groaning with pleasure and glowing and beaming with happiness. See the trailer for the film at https://vimeo.com/157202223.

26. Another version of the YWF trailer is available at http://yeswefuck.org/.

27. See Manuela Heim, "'Ich habe kein Problem damit, als Fetisch betrachtet zu werden,'" *taz*, October 18, 2014; Doris Schneider, "Koblenz: Hat Bayerlein zu viel über Sex geschrieben?—Nachfolge des Behindertenbeauftragten wird ausgeschrieben," *Rhein-Zeitung*, December 10, 2014; Stephanie Mersmann, "Koblenz: Behindertenbeauftragter Christian Bayerlein will nicht von sich aus zurücktreten," *Rhein-Zeitung*, January 12, 2015.

28. Shildrick and Price, "Deleuzian Connections."

29. Kristeva, "Liberty," 266, emphasis in original. The original French version of the essay was published in Julia Kristeva, *La Haine et le pardon: Pouvoirs et limites de la psychanalyse III* (Paris: Fayard, 2005).

30. Kristeva, "Liberty," 251, 257–58, 265.

31. Kristeva, "Liberty," 259; see Denis Diderot, *Lettre sur les aveugles à l'usage de ceux qui voient* (1749).

32. Kristeva, "Liberty," 264, 266. For the substantial improvements between 2005 and 2010 in France's provisions for persons with disabilities and especially of education for children with special needs, see European Agency for Special Needs and Inclusive Education, "France—Overview," https://

www.european-agency.org/country-information/france/national-overview
/complete-national-overview.

33. See Inclusion Europe, "The UN Convention on the Rights of Persons
with Disabilities," http://inclusion-europe.eu/?page_id=150.

34. In fact, the convention is situated in some tension between these positions; it is good that the convention takes a both/and approach, "recognizing
that disability is an evolving concept and that disability results from the interaction between persons with impairments and attitudinal and environmental
barriers that hinders their full and effective participation in society on an equal
basis with others" (Convention on the Rights of Persons with Disabilities, http://
www.un.org/disabilities/convention/conventionfull.shtml).

35. I thank Monika Baar for sharing with me documents from her work
"Rethinking Disability: The Impact of the International Year of Disabled
Persons (1981) in Global Perspective," research project, University of Leiden,
2016, http://www.hum.leiden.edu/history/research/projects-pcni/erc-rethinking
-disability.html.

36. Theresia Degener, "Die UN-Behindertenrechtskonvention: Grund-
lage für eine neue inklusive Menschenrechtstheorie," *Vereinte Nationen* 2
(2010): 57–63.

37. "Radikal normal," *Der Tagesspiegel,* December 4, 2014, http://www.tages
spiegel.de/themen/reportage/inklusionsvorreiterin-theresia-degener-im-por
traet-radikal-normal/11064492.html.

38. Degener, "Die UN-Behindertenrechtskonvention," 57, 63.

39. Theresia Degener, "Welche legislativen Herausforderungen bestehen in
Bezug auf die nationale Implementierung der UN-Behindertenrechtskonvention
in Bund und Ländern," *Behindertenrecht* 2 (2009), http://isl-ev.de/attachments
/article/910/Theresia%20Degener%20-%20Legislative%20Herausforderungen
.pdf.

40. Graumann, *Assistierte Freiheit.* Notably, already in 1998, the German
Pro Familia also advocated for assistance for individuals with cognitive disabilities who wanted to have and raise children of their own, mentioning the
existence of special projects for "accompanied parenting" (*Begleitete Eltern-
schaft*). *Sexualität und geistige Behinderung,* 15.

41. See on this point "Lohnarbeit für Selbstbestimmung? Arbeitnehmer in
der Persönlichen Behindertenassistenz," *LabourNet.de Germany,* December 18,
2012, http://archiv.labournet.de/branchen/dienstleistung/gw/cubela1.html;
and Evelyn Nakano Glenn, *Forced to Care: Coercion and Caregiving in America*
(Cambridge, MA: Harvard University Press, 2012).

42. FRA, "The Fundamental Rights." Many thanks to Robert Pursche for first researching this issue and for his great insights into the centrality of this topic for thinking about political subjectivity more generally.

43. Aichele quoted in Annelie Kaufmann, "Wählen—mit ein bisschen Hilfe: Menschen mit Behinderung dürfen nicht wählen, wenn sie in allen Belangen einen rechtlichen Betreuer haben. Menschenrechtsverbände fordern, dass sich das ändert," *Die Zeit*, June 6, 2013. See also Leander Palleit, *Gleiches Wahlrecht für alle? Menschen mit Behinderungen und das Wahlrecht in Deutschland* (Berlin: Deutsches Institut fuer Menschenrechte, 2011).

44. Terrific suggestions for how representation in financial matters could be extended also to support in political participation are made by Martha Nussbaum, "The Capabilities of People with Cognitive Disabilities," *Metaphilosophy* 40, nos. 3–4 (July 2009).

45. Article 29 of the convention calls for equality in participation in political life and mentions that assistance in the voting booth by a trusted person of choice is an option for those who need it due to disability. The challenge now is to extend this option beyond those who are physically disabled, for example, with vision issues, to those with cognitive or emotional disabilities. One argument put forward, for instance, by Verena Bentele, the current disability rights spokeswoman for the German government who is blind and a Paralympic ski champion, is that it would be terrible for Germany to trail the other, more enlightened nations: "It cannot be that with regard to political participation we in our nation are lagging behind." The ten thousand number and Bentele quoted in "Volles Wahlrecht für Behinderte gefordert," *Die Zeit*, May 5, 2014. Bentele has been the government representative for disability issues since January 2014. Further arguments are, simply, that the German situation "contradicts the goals of the UN Convention." As Valentin Aichele puts it, "Restricting this right constitutes unequal treatment, and that is not legally permissible." See Kaufmann, "Wählen."

46. FRA, "The Right to Political Participation of Persons with Mental Health Problems and Persons with Intellectual Disabilities" (Vienna, 2010), 19–20, http://fra.europa.eu/sites/default/files/fra_uploads/1216-Report-vote-dis ability_EN.pdf.

47. The publications are available at http://fra.europa.eu/en/publica tions-and-resources/publications?title=&year[min][year]=&year[max] [year]=&related_content=&field_fra_publication_type_tid_i18n[0]=86&lan guage=All.

48. Prior—albeit less deliberately antihierarchical—lifesharing experiments

had existed since the nineteenth century; the most famous of these is the v. Bodelschwinghian Foundation of Bethel (in Bielefeld), begun in 1867 to provide care for individuals with epilepsy. For the best introduction to the intensely complicated history of the Bethel institution (today the largest service organization for care for the disabled in Europe), whose director in the 1930s–1940s, Friedrich von Bodelschwingh, became an extraordinarily important figure in protesting the "euthanasia" murders and was long a signal character in the proud lore of Protestant resistance against the Nazis, see Matthias Benad, "Bethels Verhältnis zum Nationalsozialismus," in *Zwangsverpflichtet: Kriegsgefangene und zivile Zwangsarbeiter(-innen) in Bethel und Lobetal 1939–1945*, ed. Matthias Benad and Regina Mentner (Bielefeld: Verlag für Regionalgeschichte, 2002), 27–66. On the organization today, see https://www.bethel.de/startseite .html. It was only in the late 1990s and early 2000s that the more intricate story of "broken resistance" (Uwe Kaminsky's term) was explored, in which the state-loyal, conservative-nationalist views of von Bodelschwingh came to light (along with the overt sympathies for and advocacy for National Socialism of many of his coworkers). His courage in direct protest with government officials and his ability to delay deportations and protect many of the individuals in his care were also contextualized within the larger dynamics of Protestant Church and welfare organization interactions with the regime. For Uwe Kaminsky and Traugott Jähnichen's new project on daily life in Bethel in the 1920s–1940s based on patient records, see "Wie der Alltag behinderter Menschen aussah," announcement available at http://news.rub.de/presseinformationen/wissen schaft/2017-06-13-neues-projekt-wie-der-alltag-behinderter-menschen-aussah. For an excellent study of the "antagonistic cooperations" with the Nazi regime of Catholic charities providing care for the disabled during the era of "euthanasia" murders, see Winfried Süss, "Antagonistische Kooperationen: Katholische Kirche und nationalsozialistisches Gesundheitswesen in den Kriegsjahren 1939–1945," in *Kirchen im Krieg: Europa 1939–1945*, ed. Karl-Joseph Hummel and Christoph Kösters (Paderborn: Schöningh, 2009), 317–41.

    49. The history is discussed in detail and König is quoted in Zoë Brennan-Crohn, "In the Nearness of Our Striving: Camphill Communities Re-imagining Disability and Society" (BA thesis, Brown University, 2009), 28–29. Certainly, König also had his wacky views, either time-bound or purely idiosyncratic, for example, that Down syndrome was the result of a shock to the mother during pregnancy. Still, and crucially, as Brennan-Crohn points out, the alternatives of the era were dreadful, and König stood out in his refusal to accept that the children were ineducable and in his search to interact with them "in the most

meaningful and supportive way" (65). For more on Camphill, see also Dan McKanan, *Touching the World: Christian Communities Transforming Society* (Collegeville, MN: Order of Saint Benedict, 2007).

50. Brennan-Crohn, "In the Nearness," 22, 33.

51. On eugenics and conditions for the disabled in 1930s Britain, see Deborah Cohen, "Children Who Disappeared," in *Family Secrets: Privacy and Shame in Modern Britain* (New York: Oxford University Press, 2013); and Mathew Thomson, *The Problem of Mental Deficiency: Eugenics, Democracy, and Social Policy in Britain c.1870–1959* (Oxford: Clarendon Press, 1998).

52. Brennan-Crohn, "In the Nearness," 34, 46.

53. See Camphill's website, http://camphill.net/.

54. Leslie Scrivener, "Canada's Disciple to the Disabled: Jean Vanier Devotes His Life to Tearing Down Walls That Confine the Afflicted," *Presbyterian Record* (1999), https://www.thefreelibrary.com/Canada's+disciple+to+the+disabled%3A+Jean+Vanier+devotes+his+life+to...-a030090566.

55. On the theological trends, see Ulrich Bach, *Ohne die Schwächsten ist die Kirche nicht ganz: Bausteine einer Theologie nach Hadamar* (Neukirchen: Neukirchener Verlag, 2006); Amos Yong, *Theology and Down Syndrome: Reimagining Disability in Late Modernity* (Waco: Baylor University Press, 2007); John Swinton, *Becoming Friends of Time: Disability, Timefullness, and Gentle Discipleship* (Waco: Baylor University Press, 2016); Judith Z. Abrams and William C. Gaventa, eds., *Jewish Perspectives on Theology and the Human Experience of Disability* (New York: Routledge, 2006); Shelly Christensen, "Inspired by Moses: Disability and Inclusion in the Jewish Community," *Tikkun*, October 20, 2014; Hurisa Guvercin, "People with Disabilities from an Islamic Perspective," *Fountain* 63, May–June 2008, http://www.fountainmagazine.com/Issue /detail/People-with-Disabilities-from-an-Islamic-Perspective; Saulat Pervez, "The Treatment of Handicapped People in Islam," *Why Islam?*, December 13, 2014, https://www.whyislam.org/social-values-in-islam/disability-in-islam/; and Amel Alawami, "Disability in Islam," *Disabilityinfo.org*, May 31, 2017, https://blog.disabilityinfo.org/?p=5086. The classic text launching intensive theological discussion was Nancy Eiesland, *The Disabled God: Toward a Liberatory Theology of Disability* (Nashville: Abingdon Press, 1994).

56. A movie about Vanier and L'Arche released in 2017, *Summer in the Forest*, makes this quite explicit. The film begins its trailer with the remark that "some extraordinary people have a secret to reveal." It continues with Vanier commenting about the people with whom he has now shared decades of life: "They're not seeking power, they're seeking friendship. . . . It's a message for all

of us, it's about all of us." If to be a human being is to seek power, Vanier observes, "then we'll kill each other." Simultaneously, the trailer pans across Palestinian and Jewish Israeli individuals with disability enjoying each other's company. The trailer is available at https://www.youtube.com/watch?v=wLf Irx2RPrE. See also a glowing review: Mary Wakefield, "Jean Vanier's World of Love and Kindness," *Spectator*, July 1, 2017, https://www.spectator.co.uk/2017/07/jean-vaniers-world-of-love-and-kindness/.

    57. See the L'Arche Internationale website, http://www.larche.org.

    58. For example, see Christian Mahéas, "Sambo, j'ai besoin de toi," L'Arche en France, http://www.arche-france.org/temoignages/sambo-jai-besoin-toi.

    59. Data from January 2016; see the L'Arche international map at https://www.arche-deutschland.de/fileadmin/arche_national/Schriftliche_Unterlagen/AI_UEbersicht_weltweit_Stand_16-2-en.pdf (accessed May 31, 2018); see also the statistics at https://www.larche.org/in-the-world#all.

    60. On Tosquelles, Oury, and the alternative psychiatric practices, see Camille Robcis, "François Tosquelles and the Psychiatric Revolution in Postwar France," *Constellations* 23, no. 2 (June 2016): 212–22; and the interview with Robcis conducted by Todd Meyers, "Jean Oury and Clinique de La Borde: A Conversation with Camille Robcis," *Somatosphere*, June 3, 2014, http://somatosphere.net/2014/06/jean-oury-and-clinique-de-la-borde-a-conversation-with-camille-robcis.html.

    61. In Leon Hilton's beautiful summary:

> The tracings soon became a central aspect of the group's activities, and the maps grew steadily more detailed and elaborate. They developed visual systems for designating the various sounds and gestures encountered along their pathways, and started to use transparent wax paper to trace the children's daily routes. No attempt was made to interfere with their movements, or to explain or interpret them. The focus remained on the process of tracing itself. Yet distinct patterns began to emerge: certain trajectories tended to be repeated from one day to the next, and Deligny noted that some of the wandering lines seem to correspond to the conduits of underground waterways. . . . The concept of the wander line is the most significant and original contribution of Deligny's thought: it condenses, in a single stroke, his lifelong pursuit of "draining off stagnant humanisms" by unsettling the primacy of speech. He undertook the process of mapping the lines "in order to make something other than a sign." Before phrases, words, and letters can form, there must first be lines. Tracing the quotidian trajectories of his autistic collaborators, it seems, was an attempt to return to writing's origins, before it became codified or standardized, and

when it still resembled the outlines of things encountered in moving through the world. And "Deligny's language continuously evokes a kind of bodily letting go—an attenuation of subjective agency and conscious intentionality, as when one surrenders to a powerful ocean current. This quality is central to what Deligny is trying to evoke with the *lignes d'erre*, which seem to register an epistemological slackening of the distinction between the human subject and the nonhuman forces it encounters in a given environment" (Leon Hilton, "Mapping the Wander Lines: The Quiet Revelations of Fernand Deligny," *Los Angeles Review of Books*, July 2, 2015). Or as Noura Wedell argued:

> This was a materialist practice pushed to its logical conclusions. The life that took place on the raft was an elemental, material form of life, as close to objects and bare necessities as possible, a form of Paleolithic life, or life of the species: cooking, baking bread, constructing a hut for protection from summer storms, building the camp, making a fire. None of the children's doings were intentional acts in view of any functional purpose. On the contrary, they were freed of that compulsion: a child might take a basket, not to carry something, but to hang it on a hand, and keep it there for hours. . . . Deligny called such baskets wild, meaning emancipated, delivered, freed from being only what they were for those who would ascribe some fixed purpose to them. At times the children might find themselves entering into functional acts as an extension of these doings, handing a potato over to an adult who was peeling and cutting them for dinner, for example. But function was never Deligny's purpose. He was trying to find "the *detours of acting* that could allow *doing* to exist other than as a simple add-on," to be excluded and pathologized.

Noura Wedell, "To Hold a Wild Basket," *The Enemy* (2014–15), http://the enemyreader.org/to-hold-a-wild-basket/.

62. See "Le cinema de Fernand Deligny," http://www.dvdclassik.com/cri tique/le-moindre-geste-daniel-deligny-manenti.

63. See Fernand Deligny, *The Arachnean and Other Texts* (Minneapolis: Univocal, 2015); and Damian Milton, "Tracing the Influence of Fernand Deligny on Autism Studies," *Disability and Society* 31, no. 2 (2016): 285–89.

64. Hilton, "Mapping." Hilton also notes: "Writing toward the kind of affirmative accounts of autism developed by proponents of neurodiversity, Ogilvie argues that Deligny understood 'it is not on the side of autism that one finds wildness [*sauvagerie*], but rather in civilization and in its most characteristic gestures.'" See on this point also Milton, "Tracing."

65. Some examples of interesting but all too often also problematic psychoanalytic approaches to cognitive disability are Maud Mannoni, *The Backward Child and His Mother: A Psychoanalytic Study* (1964; New York: Pantheon, 1972); Frances Tustin, *Autism and Childhood Psychosis* (Hogarth Press, 1972); Valerie Sinason, *Mental Handicap and the Human Condition: An Analytic Approach to Intellectual Disability* (London: Free Association Books, 1992); Dietmut Niedecken, *Nameless: Understanding Learning Disability* (1989; New York: Brunner-Routledge, 2003). The most sensitive is the recently translated book by Angelo Villa, *Psychoanalysis and Severe Handicap: The Hand in the Cap* (2008; London: Karnac, 2013). For a somewhat ambivalent appreciation of Deligny from a currently practicing psychoanalyst's perspective, see Erik Porge, "Fernand Deligny, un style de vie avec les autistes: *Y être entre les lignes*," *Enfances & psy* 3, no. 48 (2010): 130–36.

66. Deligny, "The Arachnean," in *The Arachnean*, 4. Moreover, and note the oblique reference to Nazism: "If I wanted to indicate one of the constants of the network, I would note an 'out-side' as one of the necessary components. That said, and when space becomes a concentration camp, the formation of a network creates a kind of outside that allows the human to survive" (5).

67. Gilles Deleuze and Félix Guattari, *Anti-Oedipus: Capitalism and Schizophrenia* (1972; repr., Minneapolis: University of Minnesota Press, 1983); and Gilles Deleuze and Félix Guattari, *A Thousand Plateaus: Capitalism and Schizophrenia* (1980; repr., Minneapolis: University of Minnesota Press, 1987). It makes sense that the artist Raoef Mamedov, introduced in the preface to this book and creator of *The Last Supper* with Jesus and the disciples endowed with Down syndrome, is inspired by Deleuze and Guattari. See Ulker Mehdieva, "Rauf Mamedov: Nastoyaschiy hudozhnik sublimiruet strah smerti" [Raoef Mamedov: A true artist sublimates the fear of death], Novosti-Azerbaijan, Baku, July 29, 2013, http://ia-centr.ru/expert/16239/; translation thanks to Yuliya Barycheuskaya.

68. For these and other points on Guattari, see Dagmar Herzog, "Exploding Oedipus," in *Cold War Freud: Psychoanalysis in an Age of Catastrophes* (Cambridge: Cambridge University Press, 2016).

69. Shildrick and Price, "Deleuzian Connections."

70. Griet Roets and Dan Goodley, "Disability, Citizenship and Uncivilized Society: The Smooth and Nomadic Qualities of Self-Advocacy," *Disability Studies Quarterly* 28, no. 4 (2008), http://dsq-sds.org/article/view/131/131. What Roets and Goodley take from Deleuze and Guattari is above all an interest in

the mutual support, interdependence, and interconnection evinced by their subjects.

71. Daniela Mercieca and Duncan Mercieca, "Opening Research to Intensities: Rethinking Disability Research with Deleuze and Guattari," *Journal of Philosophy of Education* 44 (2010): 79–80, 83, 97, 90. For a marvelous related exposition of the mutual entanglement of disabled and nondisabled individuals, see Brendan Hart, "Autism Parents and Neurodiversity: Radical Translation, Joint Embodiment and the Prosthetic Environment," *BioSocieties* 9, no. 3 (September 2014): 284–303.

72. Duncan P. Mercieca, *Living* Other*wise: Students with Profound and Multiple Learning Disabilities as Agents in Educational Contexts* (Rotterdam: Sense Publishers, 2013). Like so many others who work in and with disability, Duncan Mercieca comments that we all "depend upon and receive more from other people than is commonly acknowledged," but that it is disability that brings this insight to us.

73. Cover of Matt Brim and Amin Ghaziani, eds., "Queer Methods," special issue of *WSQ* 44, nos. 3–4 (Fall–Winter 2016). The full poem is inside: Eileen Myles, "Epic for You" (265–66).

Index

George L. Mosse Series in
Modern European Cultural and Intellectual History

Series Editors

Steven E. Aschheim, Skye Doney, Mary Louise Roberts, and David J. Sorkin

Jews and Other Germans: Civil Society, Religious Diversity, and Urban Politics in Breslau, 1860–1925
Till van Rahden;
translated by Marcus Brainard

An Uncompromising Generation: The Nazi Leadership of the Reich Security Main Office
Michael Wildt;
translated by Tom Lampert

CPSIA information can be obtained
at www.ICGtesting.com
Printed in the USA
BVHW041222281118
534217BV00008B/104/P

9 780299 319205